HENRY M. MUSAL

Microwave Magnetics

MICROWAVE MAGNETICS

Ronald F. Soohoo

University of California, Davis

1817

HARPER & ROW PUBLISHERS, New York

Cambridge, Philadelphia, San Francisco,
London, Mexico City, São Paulo, Singapore, Sydney

to my wife, Rosie

Sponsoring Editor: John Willig
Project Editor: Jonathan Haber
Cover Design: Caliber Design Planning, Inc.
Text Art: Vantage Art, Inc.
Production: Delia Tedoff
Compositor: Syntax International Pte. Ltd.
Printer and Binder: The Maple Press Company

Microwave Magnetics

Library of Congress Cataloging in Publication Data

Soohoo, Ronald F.
  Microwave magnetics.

  Bibliography: p.
  Includes index.
  1. Microwave devices.   2. Magnetic materials.
I. Title.
TK7876.S66   1985   621.381′3   84-10783
ISBN 0-06-046367-8

84 85 86 87 9 8 7 6 5 4 3 2 1

Contents

Preface

This book is an introduction to the study of microwave magnetics, including relevant theory, experiments, applications, and problems. The material is based in part on notes for a course taken by seniors and graduate students at the University of California at Davis, in part on my many years of research and consulting experience. It is to a large extent self-contained so that it can be used not only as a text for students, but also for self-study by practicing engineers. Problems are provided at the end of each chapter to further the reader's understanding and application of the text material.

In recent years very few books have been written on magnetism and still fewer on the many aspects of microwave magnetics. To my knowledge, this is the first text that treats the totality of microwave magnetics in a self-contained manner. It emphasizes fundamentals rather than details to facilitate the student's understanding of the subject area.

The chapters can be divided into four groups. First, Chapters 1–3 treat the theory of microwaves. Chapter 1 introduces microwave magnetics, and Chapters 2 and 3, respectively, turn to Maxwell's equations and to waveguides and cavities.

Next, in Chapters 4 and 5, we present the fundamental theory of magnetism in order to describe the behavior of electron spins in static and time-varying electromagnetic fields. Chapter 4 classifies the different kinds of magnetism and introduces the concepts of quantum-mechanical exchange and magnetic domains. Chapter 5 analyzes the behavior of an isolated electron spin as well as that of an electron spin assembly. This allows us to formulate the equation of motion for the magnetization (defined as the magnetic moment per unit volume).

Chapters 6–9, the third group, analyze resonance behavior in various kinds of magnetic materials. Here paramagnetism, ferromagnetism, antiferromagnetism, and ferrimagnetism are discussed, along with their application to actual microwave magnetics devices. Chapter 6 is on paramagnetic resonance, including saturation effects and masers. Chapter 7 treats ferro- and antiferromagnetic resonance, including a discussion of spin waves and millimeter wave isolators employing antiferromagnetic materials. Chapter 8 concerns resonance phenomena in ferrimagnetic materials, such as ferrites and garnets; the preparation of these materials and their frequency and power limitations are also studied. Chapter 9 analyzes the design and characteristics of isolators, digital phase shifters, and junction circulators and their application in actual microwave systems.

Finally, Chapter 10 examines superconductivity, including the phenomena of single-particle and pair tunneling. The latter phenomenon, known as the Josephson effect, is a topic of current interest and emerging importance; indeed, millimeter parametric amplifiers using a Josephson junction as a variable coupling inductor have already been built.

In preparing an extensive treatise of this type, I became indebted to many scientists and engineers in the field of microwave magnetics for their original calculations, experiments, and inventions. Of course, only part of the material covered here is new, but the style and the manner of presentation are entirely my own. I am also indebted to my wife; were it not for her perserverance, understanding, and encouragement, this book would never have been completed.

Ronald F. Soohoo

Chapter 1

Introduction

In this introductory chapter we shall briefly examine the interaction between electromagnetic fields and the magnetic moment of electrons. Areas of research in microwave magnetics, as well as the application of microwave magnetics devices, will also be briefly discussed.

1.1 SCOPE OF MICROWAVE MAGNETICS

Broadly defined, microwave magnetics concerns the interaction of electromagnetic fields with the spin and, to a much lesser extent, the charge of electrons. There is an underlying unity in the subject matter of microwave magnetics when viewed in this light, despite the large number of phenomena and associated devices based on field–electron-spin interactions. Indeed, only a few basic equations are needed to facilitate the understanding of the behavior of magnetic materials and devices at microwave frequencies. For example, a problem in microwave ferrimagnetic devices involves in general the simultaneous solution of Maxwell's equations and the equation of motion of the magnetization, subject to the appropriate electromagnetic and exchange boundary conditions.

We must not surmise from this underlying simplicity, however, that the solution of these problems is necessarily simple. In fact, the geometry of most microwave magnetics devices is such that exact solutions are far from feasible.

Simplifying approximations based on physical understanding must be made if these problems are to be solved analytically. Since the analytical solution is usually a complicated function of material and device parameters, a digital computer or repeated experimentation is still needed to determine parameter values for optimum device performance.

A number of microwave magnetics devices, such as masers, ferrite isolators and circulators, and microwave superconductor devices, will be studied in this book, not as ends in themselves but as illustrations of basic principles. Nonetheless, we must keep the basic engineering objective in mind when complicated calculations are discussed lest they become solely exercises in applied mathematics.

1.2 INTERACTION OF ELECTROMAGNETIC FIELDS WITH ELECTRON ORBIT AND SPIN

To study the field-electron interaction, let us begin with Maxwell's equations. In these equations the electromagnetic fields are given in terms of the currents and charges.[1] In MKS units, Maxwell's equations are

$$\mathbf{\nabla} \times \mathbf{E} = -\frac{\partial \mathbf{B}}{\partial t} \tag{1.1}$$

$$\mathbf{\nabla} \times \mathbf{H} = \mathbf{i} + \frac{\partial \mathbf{D}}{\partial t} \tag{1.2}$$

$$\mathbf{\nabla} \cdot \mathbf{D} = \rho \tag{1.3}$$

$$\mathbf{\nabla} \cdot \mathbf{B} = 0 \tag{1.4}$$

where $\mathbf{E}$ and $\mathbf{D}$ are the electric field and electric displacement and $\mathbf{H}$ and $\mathbf{B}$ the magnetic field and magnetic flux density, respectively, $\mathbf{i}$ is the area current density, and ρ is the charge density. Equations (1.1)–(1.4) are partial differential equations since $\mathbf{E}$, $\mathbf{D}$, $\mathbf{H}$, $\mathbf{B}$, $\mathbf{i}$, and ρ may vary simultaneously with space and time. Before Maxwell's equations can be solved, they must be supplemented by another set of equations relating the quantity $\mathbf{D}$ to $\mathbf{E}$ and $\mathbf{B}$ to $\mathbf{H}$:

$$\mathbf{D} = \epsilon \mathbf{E} \tag{1.5}$$

$$\mathbf{B} = \mu \mathbf{H} \tag{1.6}$$

where ϵ is the dielectric constant and μ is the permeability.

For certain media, such as plasmas in a magnetic field, ϵ may be a tensor rather than a scalar. In that case

$$\mathbf{D} = \|\epsilon\| \mathbf{E} \tag{1.7}$$

---

[1]All *equations* in this book will be written in MKS units, consistent with current electrical engineering practice. Unfortunately, however, it is also common to measure and state *numerical values* of magnetic quantities, such as the magnetic field and magnetization, in Gaussian units. The practice of giving *values* in Gaussian units will also be followed in this text. The relationship between MKS and Gaussian units is explained in the Appendix.

where $\|\epsilon\|$ is the dielectric tensor. Similarly, for magnetic materials such as ferrites

$$\mathbf{B} = \|\mu\|\mathbf{H} \tag{1.8}$$

where $\|\mu\|$ is the permeability tensor.

If ϵ and μ are known by either experimental measurement or theoretical computation, then Eqs. (1.1)–(1.4) can be solved with the aid of Eqs. (1.5) and (1.6) or Eqs. (1.7) and (1.8).

The value of the permeability is determined mainly by the orbital and spin angular momenta of the electrons in the material. An electron may execute some orbital motion, and it always possesses an intrinsic spin about its own axis. Consider first the case of orbital motion. If an electron of charge $-e$ executes an orbital motion with frequency ω, then the corresponding current I is given by

$$I = -e\left(\frac{\omega}{2\pi}\right) \tag{1.9}$$

where ω depends on the environment to which the electron is subjected. For example, the electron may be circulating about a nucleus and at the same time subjected to a static magnetic field $\mathbf{H}$. The orbital magnetic moment μ_{or} is simply equal to I times A, where A is area of the projection of the orbital loop in the direction perpendicular to $\mathbf{H}$, and the susceptibility χ_{or} is defined as the magnetic moment per unit field. Thus, the permeability due to the orbital motion of an electron is given by

$$(\mu_{or})_{elec} = \mu_0(1 + \chi_{or}) = \mu_0\left[1 - e\left(\frac{\omega}{2\pi}\right)\frac{A}{H}\right] \tag{1.10}$$

In addition to its possible orbital motion, an electron also spins on its own axis. Since an electron has a finite mass, this spinning motion gives rise to an angular momentum that turns out to be equal to $s\hbar$ where $\hbar$ is Planck's constant divided by 2π and s is the spin. This spin angular momentum in turn gives rise to a spin magnetic moment μ_s given by

$$\mu_s = \gamma_e s\hbar \tag{1.11}$$

where the constant of proportionality $\gamma_e = -g(\mu_0 e/2m)$. Here g, called the Landé g-factor, is 2 for a free electron (apart from minor corrections arising from the quantization of the electromagnetic field) but can deviate somewhat from this value for electrons in a solid. Since γ_e is negative, μ_s and s are oppositely directed.

Consider now the application of a static magnetic field $\mathbf{H} = \hat{z}H_0$ to a free electron, $\hat{z}$ being the unit vector in the z direction. According to quantum mechanics, the projection of the spin s along $\mathbf{H}$ is either $+\frac{1}{2}$ or $-\frac{1}{2}$. Therefore, since the energy of interaction between μ_s and $\mathbf{H}$ is given by $-\mu_s \cdot \mathbf{H}$, it follows that the energy difference ΔE between the $s = \frac{1}{2}$ and $s = -\frac{1}{2}$ states is

$$\Delta E = \hbar|\gamma_e|H_0 \tag{1.12}$$

where we used Eq. (1.11). In other words, if the energy of the electron is E_0 for $H_0 = 0$, then in the presence of **H** this energy level is split into two levels, one $\Delta E/2$ above E_0 and another $\Delta E/2$ below E_0. This phenomenon is known as Zeeman splitting.

1.3 INTERACTION OF ELECTROMAGNETIC FIELDS WITH MAGNETIC MATERIALS

Consider now the application of a static magnetic field $\mathbf{H} = \hat{z}H_0$ to a material containing N spins per unit volume. Soon after the application of **H**, damping forces in the material will cause the magnetization $\mathbf{M} = N\boldsymbol{\mu}_s$ to align on the average with **H**; the alignment of **M** with **H** is opposed by random thermal motion. It follows that since the susceptibility χ_s is given by M/H_0, the spin permeability μ_{sp} of the sample is given by

$$(\mu_{sp})_{sample} = \mu_0 \left(1 + \frac{M}{H_0} \right) \tag{1.13}$$

There is, of course, also an orbital contribution to the susceptibility. If there are Z electrons per atom and N atoms per unit volume, we obtain immediately from Eq. (1.10) the susceptibility for the sample as a whole:

$$(\mu_{or})_{sample} = \mu_0 \left(1 - \frac{NZe\omega A}{2\pi H_0} \right) \tag{1.14}$$

Since the electrons in an atom can be expected to execute different orbits, A in Eq. (1.14) represents the average value of the projections of the orbital loops in the direction perpendicular to **H**. If the electrons were noninteracting, the total susceptibility would simply be the sum of the orbital and spin susceptibilities. However, in a solid, because of the presence of crystalline fields, the orbits of various electrons wobble with time so that there is on average no net orbital angular momentum in any given direction. We say in this case that the orbital angular momentum is *quenched* and that the total susceptibility is equal to μ_{sp}. However, since the quenching is usually not complete, the spin carries some orbital momentum with it. To account for this small residual effect, it is customary to use a g value nearly but not exactly equal to 2 in the expression for $\boldsymbol{\mu}_s$ given by Eq. (1.11). Since $\mathbf{M} = N\boldsymbol{\mu}_s$, the spin permeability given by Eq. (1.13) will also be slightly modified from the value corresponding to an assembly of electrons whose orbital momenta are completely quenched.

1.4 NEW AREAS OF RESEARCH IN MICROWAVE MAGNETICS

There has been a fair amount of research activity recently in the area of magnetic semiconductors. These materials, such as europium sulfide, possess simultaneously semiconducting and ferromagnetic properties. However, no application of these materials at microwave frequencies has appeared to date.

Another new area in microwave magnetics is the emission of microwave radiation by Josephson junctions formed by superconductors. This phenomenon is due to the tunneling of pairs of superconducting electrons.

Other areas of current active research include the study of magnetostatic wave delay lines, millimeter wave isolators and circulators, as well as ferrite devices for microwave integrated circuit applications.

Chapter 2

Electromagnetic Theory

In this chapter we shall study the solution to Maxwell's equations and to the wave equation, along with the derivation and application of the appropriate electromagnetic boundary conditions. The concepts of propagation constant, attenuation constant, phase constant, and wave impedance will be introduced and discussed. In addition, the concepts of electric and magnetic susceptibilities and their relation to the microscopic quantities in the Maxwell-Lorentz equations are elucidated.

2.1 MAXWELL'S EQUATIONS

2.1.1 Macroscopic Maxwell's Equations

In MKS, or practical, units, Maxwell's equations are as given in Chapter 1:

$$\nabla \times \mathbf{E} = -\frac{\partial \mathbf{B}}{\partial t} \tag{2.1}$$

$$\nabla \times \mathbf{H} = \mathbf{i} + \frac{\partial \mathbf{D}}{\partial t} \tag{2.2}$$

$$\nabla \cdot \mathbf{D} = \rho \tag{2.3}$$

$$\nabla \cdot \mathbf{B} = 0 \tag{2.4}$$

It is seen from these equations that the electromagnetic field quantities **D**, **E**, **B**, and **H** are dependent on the area current density **i** and the volume charge density ρ and vice versa. Furthermore, the relationship between the electric displacement **D** and the electric field **E** and that between the flux density

B and magnetic field **H** are given by

$$\mathbf{B} = \|\mu\|\mathbf{H} \tag{2.5}$$

$$\mathbf{D} = \|\epsilon\|\mathbf{E} \tag{2.6}$$

where the permeability $\|\mu\|$ and dielectric constant $\|\epsilon\|$ are in general second-rank tensors.[1] For example, the permeability is a tensor in the case of ferrites subjected to static and radio frequency (r.f.) magnetic fields. Likewise, the dielectric constant is a tensor for a plasma in a static magnetic field. It is not always easy to calculate $\|\mu\|$ and $\|\epsilon\|$ from first principles, although it is theoretically possible to find the expression for $\|\mu\|$ and $\|\epsilon\|$ by solving the quantum-mechanical many-body problem. In this respect, modern advances in quantum-mechanics do not invalidate Maxwell's equations but instead give a firm foundation for the calculation of $\|\mu\|$ and $\|\epsilon\|$. If $\|\mu\|$ and $\|\epsilon\|$, the components of which are usually complex, are independently measured, Eqs. (2.1)–(2.4) can be solved in a straightforward manner with the help of Eqs. (2.5) and (2.6).

In many cases the permeability and dielectric constant can be considered as scalars. For example, at sufficiently low frequencies, $\|\mu\|$ of a ferrite sample not only reduces to a scalar but is also real. At microwave frequencies the $\|\mu\|$ of the sample can also be represented by equivalent scalar permeabilities μ_n obtained from the eigenvalue equation:

$$\|\mu\|\mathbf{H}_n = \mu_n\mathbf{H}_n \tag{2.7}$$

where $\mathbf{H}_n$ are eigenfunctions belonging to the eigenvalues μ_n. A similar simplification is possible in the case of the dielectric constant. Equations (2.5) and (2.6) then simplify to

$$\mathbf{B} = \mu\mathbf{H} \tag{2.8}$$

$$\mathbf{D} = \epsilon\mathbf{E} \tag{2.9}$$

where the permeability μ represents either its low-frequency value or one of the μ_n of Eq. (2.7). To simplify the discussion, scalar permeability and dielectric constant are assumed for the rest of this chapter.

In general, μ and ϵ are not equal to their respective free-space values μ_0 and ϵ_0 because of the presence of polarization **P** and magnetization **M** in the material. In MKS units the relationships between these quantities and **B**, **H**, **D**, **E** are

$$\mathbf{B} = \mu_0(\mathbf{H} + \mathbf{M}) \tag{2.10}$$

$$\mathbf{D} = \epsilon_0\mathbf{E} + \mathbf{P} \tag{2.11}$$

It should be noted that Eqs. (2.10) and (2.11) are general relationships valid regardless of whether **M** is parallel to **H** or **P** is parallel to **E**. If **H** is not parallel to **M**, a tensor permeability results; otherwise, the permeability is a scalar. These

---

[1] In other words, $\|\mu\|$ and $\|\epsilon\|$ are 3×3 matrices having definite rules of transformation. That is, $\|\mu'\|\mathbf{H}' = \|\mu\|\mathbf{H}$ where the components of $\|\mu\|$ in the primed coordinate system are related in a definite way to those of $\|\mu\|$ in the unprimed system.

remarks are of course also applicable to the dielectric case. If we assume parallel **M** and **H** and parallel **P** and **E**, μ and ϵ can easily be deduced from Eqs. (2.10) and (2.11):

$$\mu = \frac{B}{H} = \mu_0\left(1 + \frac{M}{H}\right) = \mu_0(1 + \chi) \tag{2.12}$$

$$\epsilon = \frac{D}{E} = \epsilon_0\left(1 + \frac{P}{\epsilon_0 E}\right) = \epsilon_0(1 + \chi_e) \tag{2.13}$$

where χ and χ_e are, respectively, the magnetic and electric susceptibilities. It should be noted, however, that μ and ϵ can be defined only for linear media under static or sinusoidal conditions.

It should again be stressed that usually B, M, and H are given in Gaussian units, rather than the MKS units used in Eqs. (2.12) and (2.13) as well as in all other equations in this book. Remember that the relevant relationships between these two system of units and conversion factors between them are given in the Appendix.

2.1.2 Microscopic Maxwell Equations

It would be instructive to inquire into the atomic origin of the electromagnetic field quantities, especially for our study on the magnetic properties of solids at microwave frequencies. This discussion is based on the Maxwell-Lorentz equations, first proposed by H. A. Lorentz.[2,3] These are the microscopic counterparts of the macroscopic Maxwell equations (2.1)–(2.4).

The beginning of the twentieth century brought to light the electrical origin of matter, unknown to Maxwell when he developed his macroscopic equations in 1861–1873. This electrical origin suggests that by probing down to subatomic levels it should be possible to formulate the equations of electrodynamics in terms of charges in vacuum, that is, without the introduction of the concepts of dielectric and magnetic media. A set of microscopic field equations similar in form to the macroscopic equations (2.1)–(2.4) was proposed by Lorentz:

$$\nabla \times \mathbf{e} = -\mu_0 \frac{\partial \mathbf{h}}{\partial t} \tag{2.14}$$

$$\nabla \times \mathbf{h} = \mathbf{i}' + \epsilon_0 \frac{\partial \mathbf{e}}{\partial t} \tag{2.15}$$

$$\nabla \cdot \mathbf{e} = \frac{\rho'}{\epsilon_0} \tag{2.16}$$

$$\nabla \cdot \mathbf{h} = 0 \tag{2.17}$$

[2]H. A. Lorentz, *The Theory of Electrons*, Dover, New York, 1952.

[3]J. H. Van Vleck, *The Theory of Electric and Magnetic Susceptibilities*, Oxford University Press, New York, 1962, p. 2.

Note that μ_0 and ϵ_0 are, respectively, the permeability and dielectric constant in a vacuum. $\mathbf{i}' = \rho' \mathbf{v}'$ is the convection current density, and ρ' is the charge density with velocity $\mathbf{v}'$. The microscopic fields $\mathbf{e}$, $\mathbf{h}$ and the charge density ρ' are not the same as their macroscopic counterparts $\mathbf{E}$, $\mathbf{H}$ and ρ. Indeed, whereas Eqs. (2.14)–(2.17) are atomic or molecular in nature, Eqs. (2.1)–(2.4) are essentially statistical in nature. Hence, $\mathbf{D}$, $\mathbf{E}$, $\mathbf{B}$, $\mathbf{H}$ and ρ must be correlated in some way with the average of microscopic fields and charges over a large number of molecules.

Now let $\bar{\mathbf{e}}$ and $\bar{\mathbf{h}}$ be the average value of $\mathbf{e}$ and $\mathbf{h}$ over a physically small volume, one so small that it is inaccessible to ordinary means of measurement but nevertheless large enough to contain a very large number of molecules. It then seems reasonable, although not obvious, that the macroscopic fields $\mathbf{E}$ and $\mathbf{B}/\mu_0$ are identical to the microscopic fields $\mathbf{e}$ and $\mathbf{h}$, respectively, averaged over such a volume element[4]:

$$\mathbf{E} = \bar{\mathbf{e}} = \frac{1}{N_0} \sum_{i=1}^{N_0} \mathbf{e}_i \tag{2.18}$$

$$\mathbf{B} = \mu_0 \bar{\mathbf{h}} = \frac{\mu_0}{N_0} \sum_{i=1}^{N_0} \mathbf{h}_i \tag{2.19}$$

where $\mathbf{e}_i$ and $\mathbf{h}_i$ are the electric and magnetic fields of the ith molecule and N_0 is a large number. Similarly, the macroscopic polarization $\mathbf{P}$ and the macroscopic magnetization $\mathbf{M}$ are related to the average values of the microscopic polarization $\mathbf{p}$ and magnetization $\mathbf{m}$ as follows:

$$\mathbf{P} = N \bar{\mathbf{p}} = \frac{N}{N_0} \sum_i \mathbf{p}_i \tag{2.20}$$

$$\mathbf{M} = N \bar{\mathbf{m}} = \frac{N}{N_0} \sum_i \mathbf{m}_i \tag{2.21}$$

where $\bar{\mathbf{p}}$ and $\bar{\mathbf{m}}$ are microscopic fields over a physically small volume, as defined above, and N is the number of molecules per unit volume. It further follows that the relationships between $\mathbf{D}$, $\bar{\mathbf{e}}$, $\bar{\mathbf{p}}$ and $\mathbf{H}$, $\bar{\mathbf{h}}$, $\bar{\mathbf{m}}$ are

$$\mathbf{D} = \epsilon_0 \bar{\mathbf{e}} + N \bar{\mathbf{p}} \tag{2.22}$$

$$\mathbf{H} = \bar{\mathbf{h}} - N \bar{\mathbf{m}} \tag{2.23}$$

Equations (2.22) and (2.23) correspond to their macroscopic counterparts (2.11) and (2.10), respectively.

Averaged quantities such as $\bar{\mathbf{e}}$, $\bar{\mathbf{h}}$, $\bar{\mathbf{p}}$, and $\bar{\mathbf{m}}$ are taken for a large number of molecules at a given time. Presumably, if these quantities are to have any meaning, they must be independent of the time at which the average is taken. Such an average is called an *ensemble average*. If all molecules concerned are identical, then the ensemble average is equal to the *temporal average*, the value of a microscopic quantity such as $\mathbf{e}$ of a given molecule averaged over all times.

[4]Ibid., p. 3.

2.1.3 The Wave Equation

Equations (2.1) and (2.2) are first-order coupled equations that may be decoupled to yield second-order differential equations involving **E** or **H** only. This could be accomplished as follows. First, take the curl of both sides of Eq. (2.1) to yield

$$\mathbf{V} \times \mathbf{V} \times \mathbf{E} = -\mathbf{V} \times \frac{\partial \mathbf{B}}{\partial t} = -\frac{\partial}{\partial t}(\mathbf{V} \times \mathbf{B}) \tag{2.24}$$

The last step follows because space and time derivations are independent. We now multiply both sides of Eq. (2.2) by μ and note that $\mathbf{V} \times \mathbf{B} = \mu \mathbf{V} \times \mathbf{H}$ for an isotropic medium. Equation (2.24) becomes, with the help of Eq. (2.2),

$$\mathbf{V}(\mathbf{V} \cdot \mathbf{E}) - \nabla^2 \mathbf{E} = -\mu \frac{\partial \mathbf{i}}{\partial t} - \mu \frac{\partial^2 \mathbf{D}}{\partial t^2} \tag{2.25}$$

where we have made use of the vector identity $\mathbf{V} \times \mathbf{V} \times \mathbf{A} = \mathbf{V}(\mathbf{V} \cdot \mathbf{A}) - \nabla^2 \mathbf{A}$. In regions where $\mathbf{i}$ and ρ are zero, such as free space, Eq. (2.25) becomes

$$\nabla^2 \mathbf{E} = \mu\epsilon \frac{\partial^2 \mathbf{E}}{\partial t^2} \tag{2.26}$$

where Eq. (2.9) was used. For a plane wave, with an x component of **E** and a y component of **H**, propagating in the z direction, the solution of Eq. (2.26) is of the general form

$$E_x = C_1 f_1(z - vt) + C_2 f_2(z + vt) \tag{2.27}$$

The first term in Eq. (2.27) represents a forward traveling wave, while the second term represents a backward traveling wave. This distinction can be seen by noting that as t increases, z must also increase to keep $z - vt$ of $f_1(z - vt)$ constant, while the constancy of $f_2(z + vt)$ requires that z decreases as t increases. Because of the traveling wave character of the solution, (2.27), Eq. (2.26) is known as the *wave equation*.

In a similar manner, an equation in **H** only can be obtained by first taking the curl of Eq. (2.2) to give

$$\mathbf{V} \times \mathbf{V} \times \mathbf{H} = \mathbf{V} \times \left(\mathbf{i} + \frac{\partial \mathbf{D}}{\partial t} \right) \tag{2.28}$$

and then noting that $\mathbf{V} \times (\partial \mathbf{D}/\partial t) = \epsilon(\partial/\partial t)(\mathbf{V} \times \mathbf{E})$. Equation (2.28) becomes, with the help of Eq. (2.1),

$$\mathbf{V}(\mathbf{V} \cdot \mathbf{H}) - \nabla^2 \mathbf{H} = \mathbf{V} \times \mathbf{i} - \epsilon \frac{\partial^2 \mathbf{B}}{\partial t^2} \tag{2.29}$$

where we have again used the vector identity $\mathbf{V} \times \mathbf{V} \times \mathbf{A} = \mathbf{V}(\mathbf{V} \cdot \mathbf{A}) - \nabla^2 \mathbf{A}$. According to Eq. (2.4), $\mathbf{V}(\mathbf{V} \cdot \mathbf{H}) = 0$. In regions where $\mathbf{i} = 0$, such as free space, Eq. (2.29) reduces to

$$\nabla^2 \mathbf{H} = \mu\epsilon \frac{\partial^2 \mathbf{H}}{\partial t^2} \tag{2.30}$$

where we have used Eq. (2.5). Equation (2.30) is identical in form to Eq. (2.26); therefore its solution must be identical in form to Eq. (2.27).

Let us now return to the general wave equations (2.25) and (2.29) for $\mathbf{E}$ and $\mathbf{H}$, respectively. To obtain the solution for $\mathbf{E}$ and $\mathbf{H}$, i must be expressed in terms of $\mathbf{E}$. In a conductor i and $\mathbf{E}$ are related by Ohm's law:

$$\mathbf{i} = \sigma\mathbf{E} \tag{2.31}$$

where σ is the conductivity of the material. Using Eq. (2.31), we see that Eqs. (2.25) and (2.29) become

$$\nabla(\nabla \cdot \mathbf{E}) - \nabla^2\mathbf{E} = -\mu\sigma \frac{\partial \mathbf{E}}{\partial t} - \mu\epsilon \frac{\partial^2 \mathbf{E}}{\partial t^2} \tag{2.32}$$

$$\nabla(\nabla \cdot \mathbf{H}) - \nabla^2\mathbf{H} = -\mu\sigma \frac{\partial \mathbf{H}}{\partial t} - \mu\epsilon \frac{\partial^2 \mathbf{H}}{\partial t^2} \tag{2.33}$$

where we have again used Eqs. (2.1), (2.8), and (2.9).

Note that Eqs. (2.32) and (2.33) would be identical in form if $\nabla \cdot \mathbf{E} = \nabla \cdot \mathbf{H} = 0$ were to hold. However, although $\nabla \cdot \mathbf{B}$ is always zero—because there are *no* free magnetic poles—$\nabla \cdot \mathbf{E} = \rho/\epsilon$, which is in general nonzero, for there *are* free electric charges.[5] However, it can be shown that if any charge density existed in a conductor, it would decay extremely rapidly to zero. Taking the divergence of both sides of Eq. (2.2), we find

$$\nabla \cdot (\nabla \times \mathbf{H}) = 0 = \frac{\sigma}{\epsilon} \rho + \frac{\partial \rho}{\partial t} \tag{2.34}$$

where we have used Eqs. (2.3), (2.9), and (2.31). Assuming that the general solution of Eq. (2.34) is composed of a transient term and a steady-state term,

$$\rho = \rho_0 e^{-t/\tau} + \rho_m e^{j\omega t} \tag{2.35}$$

we readily find on substituting Eq. (2.35) into Eq. (2.34) that $\tau = \epsilon/\sigma$ and $\rho_m = 0$. For a typical metal the characteristic decay time τ is of the order of 10^{-18} s, an extremely short interval compared to the microwave period of about 10^{-10} s at X band. Thus, $\nabla \cdot \mathbf{E} = \nabla \cdot \mathbf{D}/\epsilon = \rho/\epsilon$ can be considered to be essentially zero. On the other hand, τ can be as long as 10^{-3} s in a semiconductor, much longer than a microwave period. However, in this case ρ_0 is quite small because of the rather low conductivity. Thus, $\nabla \cdot \mathbf{E} = \rho/\epsilon$ can again be neglected. Furthermore, since ρ_m is zero, the steady-state component of $\nabla \cdot \mathbf{E}$ is also zero. For these reasons, $\nabla \cdot \mathbf{E}$ as well as $\nabla \cdot \mathbf{B}$ can be set equal to zero at microwave frequencies, making Eqs. (2.32) and (2.33) identical in form.

Since Eqs. (2.32) and (2.33) are essentially identical in form, without loss of generality let us restrict ourselves to the solution of one of them, say, Eq. (2.32), for $\mathbf{E}$. Again, assume a plane wave with electric field E_x and magnetic field H_y propagating in the z direction; that is, let

$$E_x \propto e^{j\omega t - \gamma_0 z} \tag{2.36}$$

[5]Magnetic poles are not free in the sense of being separable from the magnet on which they reside. Although really free poles or monopoles have been theoretically postulated, they have not been observed.

where ω is the frequency and γ_0 is the propagation constant, a quantity to be determined. Substituting Eq. (2.36) into Eq. (2.32), we find

$$\gamma_0 = \alpha_0 + j\beta_0 = \pm\sqrt{-\omega^2\mu\left(\epsilon - j\frac{\sigma}{\omega}\right)} \tag{2.37}$$

where α_0 is the attenuation constant in nepers per meter and β_0 is the phase constant in radians per meter. In the derivation of Eq. (2.37), we note that $\nabla^2 \mathbf{E} = \hat{\mathbf{x}}\,\partial^2 E_x/\partial z^2$. If $\sigma = 0$, as in the case of a perfect insulator, Eq. (2.37) becomes

$$\gamma_0 = \alpha_0 + j\beta_0 = \pm j\omega\sqrt{\mu\epsilon} = \pm j\frac{\omega}{v} \tag{2.38}$$

where $v = 1/\sqrt{\mu\epsilon}$ is the velocity of propagation in the medium. For a poor conductor σ is small so that we can expand the square root in Eq. (2.37) using the binomial theorem and retain only terms linear in σ to obtain

$$\begin{aligned}
\gamma_0 &= \alpha_0 + j\beta_0 \\
&= \pm j\omega\sqrt{\mu\epsilon}\left(1 - j\frac{\sigma}{2\omega\epsilon} + \cdots\right) \\
&= \pm\left(\frac{\sigma}{2}\sqrt{\frac{\mu}{\epsilon}} + j\omega\sqrt{\mu\epsilon}\right)
\end{aligned} \tag{2.39}$$

On the other hand, for a good conductor (σ large) the ϵ term can be neglected compared to the σ/ω term so that Eq. (2.37) simplifies to

$$\gamma_0 = \alpha_0 + j\beta_0 = \pm\sqrt{j\omega\mu\sigma} = \pm(1+j)\sqrt{\frac{\omega\mu\sigma}{2}} \tag{2.40}$$

It is interesting to note that in this case $|\alpha_0| = |\beta_0|$. Furthermore, since $E_x \propto e^{-\gamma_0 z} = e^{-\alpha_0 z}e^{-j\beta_0 z}$, we must choose the positive sign in Eqs. (2.37)–(2.40) for waves propagating in the $+z$ direction. Conversely, the negative sign must be used for waves traveling in the $-z$ direction.

To find the relationship between $\mathbf{E}$ and $\mathbf{H}$, we note that the field quantities must satisfy Maxwell's curl equations, (2.1) and (2.2), as well as the wave equations (2.32) and (2.33). Putting it another way, only those solutions satisfying both the wave and Maxwell equations are permissible solutions. In fact, every solution that satisfies Maxwell's equations is a solution of the wave equation but not vice versa. In other words, some information is lost when the first-order coupled Maxwell's equations are uncoupled to yield the second-order wave equations. Returning now to Eq. (2.1), we have

$$-\gamma_0 E_x = -\mu j\omega H_y \tag{2.41}$$

so that a wave impedance, denoted by Z_0, can be defined as

$$Z_0 = \frac{E_x}{H_y} = \frac{j\omega\mu}{\gamma_0} = \sqrt{\frac{\mu}{\epsilon - j(\sigma/\omega)}} \tag{2.42}$$

Since Z_0, being the ratio of $|\mathbf{E}|$ to $|\mathbf{H}|$, is in general complex, there is a phase angle between the time dependence of the electric and magnetic vectors.

For a good conductor (σ large) the ϵ term can be neglected relative to the σ/ω term in Eq. (2.42). In that case

$$Z_0 = (1 + j)\sqrt{\frac{\omega\mu}{2\sigma}} \qquad (2.43)$$

It is interesting to note that as $\sigma \to \infty$, $Z_0 \to 0$. Hence, $|\mathbf{E}|/|\mathbf{H}|$ is very small in a good conductor. In contrast, $|\mathbf{E}|/|\mathbf{H}|$ for a plane wave in free space is 377 Ω.

2.2 ELECTROMAGNETIC BOUNDARY CONDITIONS

Maxwell's equations have been postulated to be applicable only for points in whose neighborhood the physical properties of the medium vary continuously. However, across any surface separating one medium from another, there occurs on the macroscopic scale sharp changes in μ, ϵ, and σ. Therefore, we may expect a corresponding occurrence of discontinuities in the electric and magnetic field quantities at these boundaries. Our aim in this section is to develop the appropriate electromagnetic boundaries conditions.

Let the surface S' shown separating medium 1 and medium 2 in Fig. 2.1 be replaced by a very thin transition layer of thickness Δl within which ϵ, μ, and σ vary rapidly but continuously from their values in medium 1 to those in medium 2. Within this layer, as within media 1 and 2, Maxwell's equations are applicable so that the field quantities and their first derivations are continuous bounded functions of coordinate and time. Thus, the introduction of a transition layer of finite thickness enables us to treat the points inside this region as ordinary points of space about which all physical quantities vary continuously. The appropriate boundary conditions can then be found by letting Δl approach 0 in our final results.

Let us draw a small right cylinder whose elements are normal to S' and whose ends lie on the surfaces of the layer so that they are separated by just

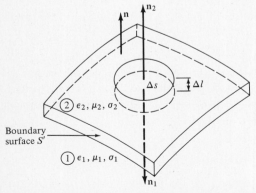

Figure 2.1 Boundary surface for determination of the normal-field-component boundary condition.

the layer thickness Δl. Since $\nabla \cdot \mathbf{B} = 0$, according to Eq. (2.4), and Gauss's theorem states that

$$\int_S \mathbf{B} \cdot \mathbf{n} \, ds = \int_V \nabla \cdot \mathbf{B} \, dv \qquad (2.44)$$

we conclude that $\int_S \mathbf{B} \cdot \mathbf{n} \, ds = 0$ when integrated over the walls and ends of the cylinder. In Eq. (2.44) S is the area while V is the volume. If the area ΔS of the ends of the cylinder is made sufficiently small, we may expect $\mathbf{B}$ to have constant values over each end of the cylinder. It then follows from Eq. (2.44) and the discussion above that

$$(\mathbf{B}_1 \cdot \mathbf{n}_1 + \mathbf{B}_2 \cdot \mathbf{n}_2) \, \Delta S + (\text{wall contribution}) = 0 \qquad (2.45)$$

with the contribution of the cylinder wall to the surface integral being directly proportional to Δl. As shown in Fig. 2.1, $\mathbf{n}_1$ and $\mathbf{n}_2$ are outward normals to the surface S'. In the limit as $\Delta l \to 0$, the transition layer will coincide with the surface S'. In this case the ends of the cylinder lie just on either side of S' so that the contribution from the cylinder wall becomes vanishingly small. Thus, as $\Delta l \to 0$ and for ΔS sufficiently small, Eq. (2.45) reduces to

$$(\mathbf{B}_2 - \mathbf{B}_1) \cdot \mathbf{n} = 0 \qquad (2.46)$$

where we have noted from Fig. 2.1 that $-\mathbf{n}_1 = \mathbf{n}_2 = \mathbf{n}$. According to Eq. (2.46), *the normal component of the magnetic flux density* $\mathbf{B}$ *across any surface of discontinuity is continuous.*

The boundary condition for the normal component of $\mathbf{D}$ may be derived in a similar manner. According to Eq. (2.3),

$$\int_V \nabla \cdot \mathbf{D} \, dv = \int_V \rho \, dv = q \qquad (2.47)$$

where q is the net charge enclosed by the volume V. With Gauss's theorem Eq. (2.47) becomes

$$\int_S \mathbf{D} \cdot \mathbf{n} \, ds = q \qquad (2.48)$$

Assume that the charge is distributed throughout the transition layer with a volume charge density ρ. It then follows from Eq. (2.47) that within the small right cylinder of Fig. 2.1,

$$q = \rho \, \Delta l \, \Delta S \qquad (2.49)$$

In the limit as $\Delta l \to 0$, so that the transition region coincides with the surface S', $\rho \to \infty$ since the total charge q within the cylinder remains constant (it cannot be destroyed). To circumvent this difficulty, let us introduce a new quantity $\Omega = \rho \Delta l$, the surface charge density in coulombs per square meter. It then follows from Eqs. (2.48) and (2.49) that

$$(\mathbf{D}_2 - \mathbf{D}_1) \cdot \mathbf{n} = \Omega \qquad (2.50)$$

It is seen from Eq. (2.50) that *the amount of discontinuity of the normal component of the electric displacement* $\mathbf{D}$ *across a boundary is equal to the surface charge density residing at that boundary.*

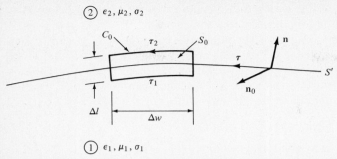

Figure 2.2 Boundary surface for determination of the tangential-field-component boundary condition.

To study the behavior of the tangential components, we replace the cylinder of Fig. 2.1 by the rectangular path of Fig. 2.2. The rectangle has an area S_0 and a perimeter C_0 with $\mathbf{n}_0$ as its outward normal. The direction of $\mathbf{n}_0$ is determined by the direction of the circulation of C_0 using the right-hand rule.

From the Stokes theorem, which states that $\int (\nabla \times \mathbf{E}) \cdot \mathbf{n}_0 \, ds = \int \mathbf{E} \cdot d\mathbf{l}$, Eq. (2.1) becomes

$$\int \mathbf{E} \cdot d\mathbf{l} = -\int \frac{\partial \mathbf{B}}{\partial t} \cdot \mathbf{n}_0 \, ds \qquad (2.51)$$

From Fig. 2.2 it then follows that if the rectangle were reasonably small,

$$(\mathbf{E}_1 \cdot \boldsymbol{\tau}_1 + \mathbf{E}_2 \cdot \boldsymbol{\tau}_2)\Delta + \text{(end contributions)} = -\frac{\partial \mathbf{B}}{\partial t} \cdot \mathbf{n}_0 \, \Delta w \, \Delta l \qquad (2.52)$$

where $\boldsymbol{\tau}_1$ and $\boldsymbol{\tau}_2$ are circulation vectors, shown in Fig. 2.2. As the transition layer contracts to the surface S', the contribution from the end segments of the rectangle, being proportional to Δl, becomes vanishingly small. From Fig. 2.2 we can define a unit tangent vector $\boldsymbol{\tau}$ by the relation

$$\boldsymbol{\tau} = \mathbf{n}_0 \times \mathbf{n} \qquad (2.53)$$

where $\mathbf{n}$ is again the positive normal to the surface S' drawn from medium 1 to medium 2. By means of vector identity $(\mathbf{n}_0 \times \mathbf{n}) \cdot \mathbf{E} = \mathbf{n}_0 \cdot (\mathbf{n} \times \mathbf{E})$, we find that as $\Delta l \to 0$ and $\Delta w \to 0$ Eq. (2.52) becomes

$$\mathbf{n}_0 \cdot \left[\mathbf{n} \times (\mathbf{E}_2 - \mathbf{E}_1) + \lim_{\Delta l \to 0} \left(\frac{\partial \mathbf{B}}{\partial t} \Delta l \right) \right] = 0 \qquad (2.54)$$

Since the orientation of the rectangle given by the direction of $\mathbf{n}_0$ is entirely arbitrary, we find

$$\mathbf{n} \times (\mathbf{E}_2 - \mathbf{E}_1) = -\lim_{\Delta l \to 0} \frac{\partial \mathbf{B}}{\partial t} \Delta l \qquad (2.55)$$

Under the usual assumption that the field vectors and their time derivatives are finite in magnitude, Eq. (2.55) reduces to

$$\mathbf{n} \times (\mathbf{E}_2 - \mathbf{E}_1) = 0 \qquad (2.56)$$

when the limit $\Delta l \to 0$ is taken. We find therefore that *the tangential component of the electric field* **E** *is continuous across the boundary.*

The boundary condition for the tangential component of **H** can be similarly deduced. Instead of starting with the curl equation (2.1), which states that $\mathbf{V} \times \mathbf{E} = -\partial \mathbf{B}/\partial t$, we should begin with Eq. (2.2), which reads $\mathbf{V} \times \mathbf{H} = \mathbf{i} + \partial \mathbf{D}/\partial t$. It is then evident that the required expression can be obtained by analogy. Specifically, change the **E**s to **H**s and $-\partial \mathbf{B}/\partial t$ to $\mathbf{i} + \partial \mathbf{D}/\partial t$ in Eq. (2.55) to obtain

$$\mathbf{n} \times (\mathbf{H}_2 - \mathbf{H}_1) = \lim_{\Delta l \to 0} \left(\mathbf{i} + \frac{\partial \mathbf{D}}{\partial t} \right) \Delta l \qquad (2.57)$$

Since $\partial \mathbf{D}/\partial t$ is assumed to be finite,

$$\lim_{\Delta l \to 0} \left(\frac{\partial \mathbf{D}}{\partial t} \Delta l \right) \to 0 \qquad \text{as} \quad \Delta l \to 0$$

If the current density **i** is also finite, then

$$\lim_{\Delta l \to 0} (\mathbf{i} \, \Delta l) \to 0 \qquad \text{as} \quad \Delta l \to 0$$

If, however, the total $I = \mathbf{i} \cdot \mathbf{n}_0 \, \Delta l \, \Delta w$ remains finite as $\Delta l \to 0$, then **i** may approach infinity. To circumvent this difficulty, let $\mathbf{K} = \mathbf{i} \, \Delta l$ be the surface current density as $\Delta l \to 0$. It then follows from Eq. (2.57) that

$$\mathbf{n} \times (\mathbf{H}_2 - \mathbf{H}_1) = \mathbf{K} \qquad (2.58)$$

Since $\mathbf{i} = \sigma \mathbf{E}$, it could only be infinite if σ were infinite since **E** is by necessity finite. Therefore, Eq. (2.58) is applicable if the conductivity of one of the mediums is infinite. This assumption has an important implication. For example, waveguide walls are usually assumed to have infinite conductivity to facilitate the solution of waveguide problems. In this case we see from Eq. (2.58) that *for infinite conductivity, the discontinuity in the magnetic field* **H** *at a boundary is equal to the surface current density at that boundary.*

On the other hand, if the conductivity is assumed finite, $\mathbf{i} = \sigma \mathbf{E}$ and therefore **K** must be zero as $\Delta l \to 0$ since **E** is by necessity finite. It follows therefore that *the magnetic field is continuous across a boundary of finite conductivity.*

PROBLEMS

2.1 (a) Starting from Maxwell's equations, derive the equation in **H** only for the case where the permittivity and conductivity are both scalars designated by ϵ and σ and permeability is a tensor $\|\mu\|$.

    (b) Find the relationship between the components of $\|\mu\|$ and **H** such that the equation obtained in (a) will reduce to a wave equation.

    (c) If $\mu_{11} = \mu_{22} = \mu_{33}$ and $\mu_{12} = \mu_{13} = \mu_{21} = \mu_{23} = \mu_{31} = \mu_{32} = 0$, what is the spatial and time dependence of **H** for the case described in (b)?

2.2 Assume that the primed and unprimed cartesian coordinate systems are at arbitrary orientation with respect to each other. If $\|\mu\|$ of the unprimed system is given by

$$\|\mu\| = \begin{bmatrix} \mu_{11} & \mu_{12} & \mu_{13} \\ \mu_{21} & \mu_{22} & \mu_{23} \\ \mu_{31} & \mu_{32} & \mu_{33} \end{bmatrix}$$

find $\|\mu\|$ of the primed system.

2.3 In vacuo, the Maxwell-Lorentz microscopic equations (2.14)–(2.17) are similar in form to the Maxwell macroscopic equations (2.1)–(2.4). However, $\mathbf{i}'$ and $\mathbf{i}$ are not the same. Enumerate the differences between them.

2.4 (a) Suppose that a capacitor partially filled with a metallic slab of conductivity σ is connected across a battery via a switch. Assuming that the switch is closed at time zero, sketch the current through the circuit as a function of time.

(b) Given that the metallic slab fills the entire capacitor, sketch again the circuit current as a function of time, assuming that the switch is closed at $t = 0$.

2.5 If a uniform magnetic field increasing linearly at a time rate k from $t = 0$ is applied perpendicular to a current-carrying circuit of impedance Z, what is the change in circuit current due to the application of the field if the effective area of the circuit loop is A?

2.6 According to the discussion connected with Eq. (2.35), the steady-state component of $\mathbf{V} \cdot \mathbf{E}$ is always zero regardless of the values of ϵ, σ, or ω. Is this conclusion consistent with the skin-depth concept? Start your discussion by deriving the skin-depth equation:

$$\nabla^2 \mathbf{H} = \mu\sigma \frac{\partial \mathbf{H}}{\partial t}$$

2.7 (a) Consider the case where the plane surface of a perfectly conducting metal is coated with a dielectric layer of dielectric constant $\epsilon (= 5\epsilon_0)$ and thickness δ. If the electric field on the air side of the air-dielectric interface has magnitude E_0 and is oriented $45°$ to the boundary, find the electric field $\mathbf{E}$ on the dielectric side of the boundary.

(b) Find the tangential component of $\mathbf{E}$ and charge density at the dielectric-metal boundary.

2.8 (a) Consider a parallel-plane transmission line made of perfect conductors lying in the y-z plane and separated by a distance d. Let the current per unit width flowing in the upper conductor be given by

$$\mathbf{I} = \hat{\mathbf{z}} I_0 e^{j(\omega t - \beta z)}$$

where $\hat{\mathbf{z}}$ is the unit vector in the z direction and β is the phase constant. Find the direction and magnitude of $\mathbf{H}$ at the inner surface of the upper and lower conductors in terms of the current.

(b) Would there be a displacement current flowing in the transmission line? Explain.

Chapter 3

Waveguides and Cavities

In this chapter we shall study the theory of waveguides (coaxial, rectangular, cylindrical, stripline, microstrip, or dielectric image), cavities, and associated quantities. These include the reflection coefficient, standing wave ratio, impedance, power flow, energy density, and the phase and group velocities. The treatment is essentially self-contained and emphasizes physical reasoning as well as mathematical rigor. This chapter should be a good review for those who have studied waveguides and cavities. For those who have not, it will serve as an introduction to the understanding of more complicated structures used in microwave electronics and magnetics.

3.1 GENERAL SOLUTION OF WAVE EQUATIONS

In Chapter 2 we derived the wave equations (2.32) and (2.33) for $\mathbf{E}$ and $\mathbf{H}$, respectively. Setting $\nabla \cdot \mathbf{E} = 0$, a condition valid for all sinusoidal cases, as explained in connection with Eq. (2.35), we find that these equations reduce to

$$\nabla^2 \mathbf{E} - \gamma_0^2 \mathbf{E} = 0 \tag{3.1}$$

$$\nabla^2 \mathbf{H} - \gamma_0^2 \mathbf{H} = 0 \tag{3.2}$$

where

$$\gamma_0^2 = -\omega^2 \mu \left(\epsilon - \frac{j\sigma}{\omega} \right)$$

is first defined by Eq. (2.37) and $\mathbf{E}$ and $\mathbf{H}$ have been assumed to have the time dependence $e^{j\omega t}$. Since Eqs. (3.1) and (3.2) are of the same form, we may concentrate on the solution of one of them, Eq. (3.1), say, without any loss of generality.

3.1.1 Rectangular Coordinates

In rectangular coordinates $\nabla^2 \mathbf{E} = \hat{\mathbf{x}} \nabla^2 E_x + \hat{\mathbf{y}} \nabla^2 E_y + \hat{\mathbf{z}} \nabla^2 E_z$. Therefore, Eq. (3.1) could be decomposed into three scalar equations that are identical in form, one for each component of $\mathbf{E}$. For example, for the z component, we have

$$\nabla^2 E_z - \gamma_0^2 E_z = 0 \tag{3.3}$$

If $\mathbf{E} = e^{j\omega t - \gamma z}$, Eq. (3.3) becomes

$$\frac{\partial^2 E_z}{\partial x^2} + \frac{\partial^2 E_z}{\partial y^2} + (\gamma^2 - \gamma_0^2)E_z = 0 \tag{3.4}$$

where γ is known as the propagation constant. Let the quantity $\gamma^2 - \gamma_0^2$ be designated k_c^2 where k_c is the waveguide-geometry-dependent cutoff wave number. Then it follows that $\gamma^2 = k_c^2 + \gamma_0^2$, a quantity whose physical significance we can readily appreciate. For example, for a lossless waveguide containing a lossless medium $\gamma_0^2 = -\omega^2\mu\epsilon$, so that

$$\gamma^2 = k_c^2 - \omega^2\mu\epsilon \tag{3.5}$$

Because $E_z \propto e^{-\gamma z}$, γ is evidently real, indicating attenuation without propagation for frequencies $\omega < k_c/\sqrt{\mu\epsilon}$. For $\omega > k_c/\sqrt{\mu\epsilon}$, γ is imaginary so that we have propagation without attenuation. In this respect, a waveguide is like a high-pass filter with a transition frequency ω_c from high loss to low loss equal to $k_c/\sqrt{\mu\epsilon}$. However, as we shall see, unlike the lumped-parameter high-pass filter, k_c may take on a number of discrete values, each for a particular waveguide mode. Associated with the cutoff (transition) frequency ω_c, there is the cutoff wavelength $\lambda_c = 2\pi/k_c$; λ_c is defined as the wavelength a disturbance of frequency ω_c would have if propagated in a plane wave. It therefore follows that λ_c and ω_c are related as $\lambda_c = 2\pi/\omega_c\sqrt{\mu\epsilon}$.

If the waveguide medium is not lossless, Eq. (3.5) must be generalized to read

$$\gamma^2 = k_c^2 - \omega^2\mu\left(\epsilon - j\frac{\sigma}{\omega}\right) \tag{3.6}$$

where we have used Eq. (2.37). It is evident from Eq. (3.6) that in this case γ is complex, indicating propagation with attenuation. If the medium inside the waveguide is slightly absorbing, the transition from high to low loss occurs gradually over a small band of frequencies rather than at a single frequency. If the loss is sufficiently large, the transition may occur over such a broad band of frequencies that the very definition of a cutoff frequency may become debatable.[1]

Letting $\nabla^2_{x,y} = \partial^2/\partial x^2 + \partial^2/\partial y^2$, Eq. (3.4) becomes

$$\nabla^2_{x,y}E_z = -k_c^2 E_z \tag{3.7}$$

Since we are mostly interested in lossless waveguiding systems, the simpler Eq. (3.5) applies. Equation (3.7) can be solved by the well-known method of separation of variables. To do this, assume that E_z can be represented by the product of two functions $X(x)$ and $Y(y)$:

$$E_z = X(x)Y(y) \tag{3.8}$$

As these notations indicate, X is a function of x only while Y is a function of y only. Substituting Eq. (3.8) into Eq. (3.7) we find

$$\frac{X''}{X} + \frac{Y''}{Y} = -k_c^2 \tag{3.9}$$

where $X'' = \partial^2 X/\partial x^2$ and $Y'' = \partial^2 Y/\partial y^2$. Since x and y are independent variables, the first and second terms of Eq. (3.9) can be independently varied. It follows that equality (3.9) holds for all values of x and y only if the ratios X''/X and Y''/Y are constants. There are several forms for the solution of Eq. (3.9), according to whether these ratios are both positive, both negative, or one positive and one negative.

If $X''/X = K_x^2$ and $Y''/Y = K_y^2$ are both positive, the solution of Eq. (3.9) is

$$\begin{aligned} X &= A_1 \cosh K_x x + B_1 \sinh K_x x \\ Y &= C_1 \cosh K_y y + D_1 \sinh K_y y \end{aligned} \tag{3.10}$$

where $K_x^2 + K_y^2 = -k_c^2$. If $X''/X = -k_x^2$ and $Y''/Y = -k_y^2$ are both negative, the solution of Eq. (3.9) is

$$\begin{aligned} X &= A_2 \cos k_x x + B_2 \sin k_x x \\ Y &= C_2 \cos k_y y + D_2 \sin k_y y \end{aligned} \tag{3.11}$$

where $-k_x^2 - k_y^2 = -k_c^2$. If $X''/X = K_x^2$ is positive and $Y''/Y = -k_y^2$ is negative, the solution of Eq. (3.9) is

$$\begin{aligned} X &= A_3 \cosh K_x x + B_3 \sinh K_x x \\ Y &= C_3 \cos k_y y + D_3 \sin k_y y \end{aligned} \tag{3.12}$$

where $K_x^2 - k_y^2 = -k_c^2$. On the other hand, if $X''/X = -k_x^2$ is negative and $Y''/Y = K_y^2$ is positive, the solution of Eq. (3.9) is

$$\begin{aligned} X &= A_4 \cos k_x x + B_4 \sin k_x x \\ Y &= C_4 \cosh K_y y + D_4 \sinh K_y y \end{aligned} \tag{3.13}$$

---

[1] If the waveguide is also assumed to be lossy, the solution for γ would be much more complicated. Furthermore, whether loss occurs in the medium or in the guidewalls, coupling between waveguide modes may occur. For related discussion see footnote 3 of this chapter.

where $-k_x^2 + K_y^2 = -k_c^2$. Since H_z satisfies an equation identical in form to Eq. (3.9), solutions for H_z are also of the same form, as given by Eqs. (3.10)–(3.13).

3.1.2 Cylindrical Coordinates

In cylindrical coordinates Eq. (3.1) becomes

$$\frac{\partial^2 E_z}{\partial r^2} + \frac{1}{r}\frac{\partial E_z}{\partial r} + \frac{1}{r^2}\frac{\partial^2 E_z}{\partial \theta^2} = -k_c^2 E_z \tag{3.14}$$

Assuming that $E_z = R(r)F_\theta(\theta)$, we find

$$R''F_\theta + \frac{R'F_\theta}{r} + \frac{F_\theta''R}{r^2} = -k_c^2 R F_\theta \tag{3.15}$$

where $R'' = \partial^2 R/\partial r^2$, $R' = \partial R/\partial r$, and $F'' = \partial^2 F/\partial \theta^2$. Dividing Eq. (3.15) through by RF_θ and multiplying through by r^2 yields

$$r^2\frac{R''}{R} + \frac{rR'}{R} + k_c^2 r^2 = -\frac{F_\theta''}{F_\theta} \tag{3.16}$$

The left-hand side of the equation is a function of r, while the right side is a function of θ only. If both sides are to be equal for all values of r and θ, they must equal a constant, say, n^2. Then

$$-F_\theta'' = n^2 F_\theta \tag{3.17}$$

and

$$r^2\frac{R''}{R} + \frac{rR'}{R} + k_c^2 r^2 = n^2 \tag{3.18}$$

or

$$R'' + \frac{R'}{r} + \left(k_c^2 - \frac{n^2}{r^2}\right)R = 0 \tag{3.19}$$

The solutions of Eq. (3.17) are sinusoids. Solutions of Eq. (3.19) are given in terms of Bessel functions J_n and Neumann functions N_n. Thus,

$$\begin{aligned} R &= AJ_n(k_c r) + BN_n(k_c r) \\ F_\theta &= C\cos n\theta + D\sin n\theta \end{aligned} \tag{3.20}$$

or

$$\begin{aligned} R &= A_1 H_n^{(1)}(k_c r) + B_1 H_n^{(2)}(k_c r) \\ F_\theta &= C\cos n\theta + D\sin n\theta \end{aligned} \tag{3.21}$$

or

$$\begin{aligned} R &= A_2 J_n(k_c r) + B_2 H_n^{(1)}(k_c r) \\ F_\theta &= C\cos n\theta + D\sin n\theta \end{aligned} \tag{3.22}$$

where

$$\begin{aligned} H_n^{(1)}(k_c r) &= J_n(k_c r) + jN_n(k_c r) \\ H_n^{(2)}(k_c r) &= J_n(k_c r) - jN_n(k_c r) \end{aligned} \tag{3.23}$$

$H_n^{(1)}$ and $H_n^{(2)}$ are called Hankel functions of the first and second kind. A few Bessel and Neumann functions are plotted in Fig. 3.1, and all the functions

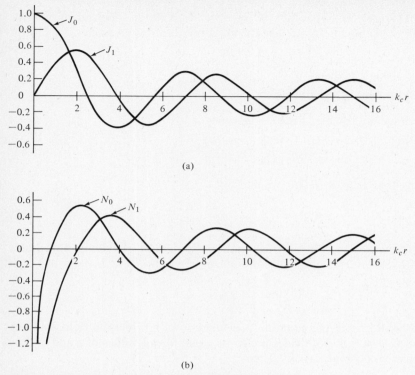

Figure 3.1 **(a)** Bessel functions and **(b)** Neumann functions.

cited in this section have been tabulated by Jahnke and Emde.[2] Since H_z satisfies an equation identical in form to Eq. (3.14), solutions for H_z are also of the same form, as given by Eqs. (3.20)–(3.23).

To recapitulate, we have derived the expressions for the longitudinal components of **E** (i.e., E_z) by solving the wave equation (3.1) by the method of separation of variables. In rectangular coordinates expressions for $E_z = XY$ are given by Eq. (3.10), (3.11), (3.12), or (3.13) depending on the signs for the ratios X''/X and Y''/Y. We pointed out previously in connection with Eq. (3.3) that the wave equation (3.1) can be decomposed into three scalar equations that are identical in form, each for one of the components of **E**. Therefore, the expressions given by Eqs. (3.10)–(3.13) for E_z should also be applicable for E_x and E_y. However, as there are four constants for each expression, the spatial distribution for E_x, E_y, and E_z may nevertheless be very different depending upon the relative values of the constants for a given problem. Furthermore, as we shall demonstrate below, these components are not independent of each other. In fact, once E_z and H_z are specified, E_x, E_y, H_x, and H_y are likewise determined. Similar remarks can be made with regard to the solutions in cylindrical coordinates.

[2]E. Jahnke and F. Emde, *Tables of Functions*, Dover, New York, 1943.

3.2 RELATIONSHIP BETWEEN LONGITUDINAL AND TRANSVERSE COMPONENTS

To find the relationship between the longitudinal and transverse components of the electromagnetic fields, we begin with Maxwell's equations (2.1) and (2.2), rewritten below:

$$\mathbf{V} \times \mathbf{E} = -\gamma_0 Z_0 \mathbf{H} \tag{3.24}$$

$$\mathbf{V} \times \mathbf{H} = \frac{\gamma_0}{Z_0} \mathbf{E} \tag{3.25}$$

where $\quad \gamma_0 = \sqrt{-\omega^2 \mu(\epsilon - j\sigma/\omega)} \quad$ and $\quad Z_0 = \sqrt{\dfrac{\mu}{(\epsilon - j\sigma/\omega)}}$

as given, respectively, by Eqs. (2.37) and (2.42). Taking the cross product of $\hat{\mathbf{z}}$ with Eq. (3.24) and using the vector identity $\mathbf{A} \times (\mathbf{V} \times \mathbf{B}) = \mathbf{V}(\mathbf{A} \cdot \mathbf{B}) - (\mathbf{A} \cdot \mathbf{V})\mathbf{B} - (\mathbf{B} \cdot \mathbf{V})\mathbf{A} - \mathbf{B} \times (\mathbf{V} \times \mathbf{A})$, we obtain

$$\mathbf{V} E_z + \gamma \mathbf{E} = -\gamma_0 Z_0 \hat{\mathbf{z}} \times \mathbf{H} \tag{3.26}$$

Similarly, taking the cross product of $\hat{\mathbf{z}}$ with Eq. (3.25) and again using the vector identity above, we find

$$\mathbf{V} H_z + \gamma \mathbf{H} = \frac{\gamma_0}{Z_0} \hat{\mathbf{z}} \times \mathbf{E} \tag{3.27}$$

Taking the cross product of $\hat{\mathbf{z}}$ with Eq. (3.27) and substituting the expression for $\hat{\mathbf{z}} \times \mathbf{H}$ obtained from Eq. (3.26) into the resulting equation, we obtain

$$\mathbf{E}_t = -\frac{\gamma}{k_c^2} \mathbf{V}_t E_z + \frac{\gamma_0 Z_0}{k_c^2} (\hat{\mathbf{z}} \times \mathbf{V}_t H_z) \tag{3.28}$$

where we have decomposed $\mathbf{E}$ into $\mathbf{E}_t + \hat{\mathbf{z}} E_z$ so that $\mathbf{E}_t$ is the component of $\mathbf{E}$ transverse to the direction of propagation. Similarly, $\mathbf{V}_t E_z = \hat{\mathbf{x}} \, \partial E_z / \partial x + \hat{\mathbf{y}} \, \partial E_z / \partial y$ is the transverse component of the gradient of E_z, etc. Similarly, taking the cross product of $\hat{\mathbf{z}}$ with Eq. (3.26) and substituting the expression for $\mathbf{z} \times \mathbf{E}$ obtained from Eq. (3.27) into the resulting equation, we obtain

$$\mathbf{H}_t = -\frac{\gamma}{k_c^2} \mathbf{V}_t H_z - \frac{\gamma_0}{k_c^2 Z_0} (\hat{\mathbf{z}} \times \mathbf{V}_t E_z) \tag{3.29}$$

Again, $\mathbf{H} = \mathbf{H}_t + \hat{\mathbf{z}} H_z$ and $\mathbf{V}_t H_z = \hat{\mathbf{x}} \, \partial H_z / \partial x + \hat{\mathbf{y}} \, \partial H_z / \partial y$, etc. Thus, we have succeeded in expressing the *transverse* components of $\mathbf{E}$ and $\mathbf{H}$ in terms of the derivatives of the *longitudinal* components, as given by Eqs. (3.28) and (3.29). Since E_z and H_z, as given by Eqs. (3.10)–(3.13) or (3.20)–(3.23), are solutions of the wave equation and Eqs. (3.28) and (3.29) were derived from Maxwell's equations, E_t, H_t, E_z, H_z satisfy both the wave equation and Maxwell's equations, as required.

3.3 TE AND TM MODES

We note from Eqs. (3.28) and (3.29) that $\mathbf{E}_t$ and $\mathbf{H}_t$ can exist if either E_z or H_z is equal to zero but not if both are zero. Accordingly, we can subdivide the permissible solutions into two types.

The first describes a wave whose longitudinal component of the electric field E_z is equal to zero; that is, the electric field $\mathbf{E}$ of such a wave lies entirely in the transverse plane. For this reason, it is customary to designate such a wave as a transverse electric, or *TE mode*. As H_z must be nonzero for such a wave lest $\mathbf{E}_t$ and $\mathbf{H}_t$ also vanish according to Eqs. (3.28) and (3.29), such a wave is sometimes called an $\mathbf{H}$ mode, a notation often used in physics.

The second solution type is a wave whose longitudinal component of the magnetic field H_z is equal to zero. Since the magnetic field $\mathbf{H}$ of such a wave must lie entirely in the transverse plane, it is designated as a transverse magnetic, or *TM mode*. Because E_z must be nonzero for such a mode, it is sometimes also called an E mode by physicists.

For $E_z = 0$ we have from Eqs. (3.28) and (3.29) that

for TE modes
$$\mathbf{E}_t = \frac{\gamma_0 Z_0}{k_c^2}\, \hat{\mathbf{z}} \times \mathbf{V}_t H_z \tag{3.30}$$

$$\mathbf{V}_t H_z = -\frac{k_c^2}{\gamma}\, \mathbf{H}_t \tag{3.31}$$

Similarly, for $H_z = 0$ we have, also from Eqs. (3.28) and (3.29), that

for TM modes
$$\mathbf{H}_t = -\frac{\gamma_0}{k_c^2 Z_0}\, (\hat{\mathbf{z}} \times \mathbf{V}_t E_z) \tag{3.32}$$

$$\mathbf{V}_t E_z = -\frac{k_c^2}{\gamma}\, \mathbf{E}_t \tag{3.33}$$

The relationship between E_t and H_t for TE modes could be found from combining Eqs. (3.30) and (3.31):

$$\mathbf{E}_t = -Z_{\text{TE}} \hat{\mathbf{z}} \times \mathbf{H}_t \tag{3.34}$$

where
$$Z_{\text{TE}} = \frac{\gamma_0}{\gamma}\, Z_0 \tag{3.35}$$

is called the wave impedance for the TE modes. Taking the cross product of $\hat{\mathbf{z}}$ and Eq. (3.34) transforms Eq. (3.34) into

$$Z_{\text{TE}} \mathbf{H}_t = \hat{\mathbf{z}} \times \mathbf{E}_t \tag{3.36}$$

Similarly, for TM modes, we can combine Eqs. (3.32) and (3.33) to get

$$Z_{\text{TM}} \mathbf{H}_t = \hat{\mathbf{z}} \times \mathbf{E}_t \tag{3.37}$$

where
$$Z_{\text{TM}} = \frac{\gamma}{\gamma_0}\, Z_0 \tag{3.38}$$

is the wave impedance of the TM modes.

We notice from Eqs. (3.36) and (3.37) that since $\hat{z}$ and $\mathbf{E}_t$ are at right angles to each other, $\mathbf{H}_t$ is at right angles to $\mathbf{E}_t$ in the x-y plane and equal in magnitude to $|\mathbf{E}_t|$ divided by the characteristic impedance Z_{TE} or Z_{TM} for the TE and TM modes, respectively. Of course, the relationships between the longitudinal and transverse components of the electromagnetic fields are given by Eqs. (3.30) and (3.31) for TE modes and by Eqs. (3.32) and (3.33) for TM modes.

In microwave magnetics problems we are usually dealing with electromagnetic fields contained in some form of metallic waveguiding structures. It is therefore pertinent to find the specific boundary conditions on E_z and H_z at a dielectric-metallic boundary. Referring to Fig. 2.1, we shall assume that medium 1 is a perfect conductor ($\sigma = \infty$) and that medium 2 is a dielectric.[3]

From Eq. (2.56) we have

$$\hat{n} \times (\mathbf{E}_2 - \mathbf{E}_1) = 0 \tag{3.39}$$

It follows that

$$\hat{n} \times \mathbf{E}_2 = \hat{n} \times \mathbf{E}_1 \tag{3.40}$$

It is clear from Eq. (3.40) that if the tangential component of $\mathbf{E}_1$ is determined, so is the tangential component of $\mathbf{E}_2$. To find $\mathbf{E}_1$, we can use Eq. (2.32), repeated below:

$$\nabla(\nabla \cdot \mathbf{E}_1) - \nabla^2 \mathbf{E}_1 = -\mu\sigma \frac{\partial \mathbf{E}_1}{\partial t} - \mu\epsilon \frac{\partial^2 \mathbf{E}_1}{\partial t^2} \tag{3.41}$$

If $\mathbf{E}_1$ has a sinusoidal dependence, we find from Eq. (2.2) that $\nabla \cdot \mathbf{E}_1 = 0$ since

$$\nabla \cdot (\nabla \times \mathbf{H}_1) \equiv 0 = (\sigma + j\omega\epsilon)\nabla \cdot \mathbf{E}_1 \tag{3.42}$$

where we have used Eqs. (2.9) and (2.31). Substituting Eq. (3.42) into Eq. (3.41), we have

$$\nabla^2 \mathbf{E}_1 = j\omega\mu(\sigma + j\omega\epsilon)\mathbf{E}_1 \tag{3.43}$$

The solution of Eq. (3.43) shows that as we go in from the surface of the metal $\mathbf{E}$ decays with a characteristic decay distance equal to the real part of $1/\sqrt{j\omega\mu(\sigma + j\omega\epsilon)}$, the distance below the surface at which $\mathbf{E}$ decays to $1/e$ of its value at the surface. The real part of $1/\sqrt{j\omega\mu(\sigma + j\omega\epsilon)}$ approaches zero as σ approaches infinity. We therefore conclude that $\mathbf{E}_1$ is zero at all points inside a perfect conductor. Accordingly, Eq. (3.40) becomes

$$\hat{n} \times \mathbf{E}_2 = \hat{n} \times \mathbf{E}_1 = 0 \tag{3.44}$$

at the surface of a perfect conductor. Since $\hat{z}E_z$ is always tangential to the waveguide wall, we find

for TM modes $E_z = 0$ at the surface of a perfect conductor $\qquad$ (3.45)

---

[3]Although the conductivity of metals used for waveguides is not infinite, it is nevertheless sufficiently large that the field distribution within the guide, assuming infinite conductivity, closely approximates that for the actual guide. For finite wall conductivity boundary conditions at both sides of the guide walls must be applied. Furthermore, single-mode propagation may no longer be assumed due to coupling between modes as a result of losses.

Now let us turn our attention to the components of **H**. From Eq. (2.46) we have

$$\hat{n} \cdot \mathbf{B}_2 = \hat{n} \cdot \mathbf{B}_1 \tag{3.46}$$

From Eq. (2.33) we againt conclude that $\mathbf{B}_1$ is zero for all points within the perfect conductor. Thus, since $\mathbf{B}_2 = \mu \mathbf{H}_2$,

$$\hat{n} \cdot \mathbf{B}_2 = \hat{n} \cdot \mathbf{H}_2 = 0 \tag{3.47}$$

Furthermore, according to Eq. (3.31),

$$\mathbf{\nabla}_t H_{z2} = -\frac{k_c^2}{\gamma} \mathbf{H}_{t2} \tag{3.48}$$

It follows that for the component normal to the conductor surface,

$$\frac{\partial H_{z2}}{\partial n} = -\frac{k_c^2}{\gamma} (\hat{n} \cdot \mathbf{H}_{t2}) \tag{3.49}$$

where $\hat{n}$ is the surface normal. Combining Eqs. (3.47) and (3.49), we find

for TE modes $\quad \dfrac{\partial H_z}{\partial n} = 0 \quad$ at the surface of a perfect conductor $\tag{3.50}$

3.4 WAVEGUIDES

3.4.1 Rectangular Waveguides

Refer now to the rectangular waveguide of the inset to Fig. 3.2. From Eq. (3.11) we have

for TM modes $\quad\quad\quad E_z = A \sin k_x x \sin k_y y \tag{3.51}$

where $A = B_2 D_2$. A_2 and C_2 have been set equal to zero since $E_z = 0$ at $x = 0$ and $E_z = 0$ at $y = 0$, according to Eq. (3.45) and Fig. 3.2. Similarly, we have from Eq. (3.11)

for TE modes $\quad\quad\quad H_z = B \cos k_x x \cos k_y y \tag{3.52}$

where $B = A_2 C_2$. B_2 and D_2 have been set equal to zero as $\partial H_z / \partial x = 0$ at $x = 0$ and $\partial H_z / \partial y = 0$ at $y = 0$, according to Eq. (3.50).

The requirement of $E_z = 0$ at $x = a$ and at $y = b$ determines the permissible values of k_x and k_y for TM modes. Accordingly, we have from Eq. (3.51) that

for TM modes $\quad\quad\quad k_x a = m\pi \quad$ or $\quad k_x = \dfrac{m\pi}{a} \tag{3.53}$

$$k_y b = n\pi \quad \text{or} \quad k_y = \frac{n\pi}{b} \tag{3.54}$$

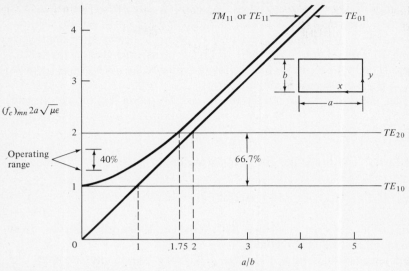

Figure 3.2   Cutoff frequency vs ratio of guide dimensions.

where m and n are integers. The requirement that $\partial H_z/\partial x = 0$ at $x = a$ and at $y = b$ leads to the *same* permissible values of k_x and k_y for TE modes as those given for TM modes in Eqs. (3.53) and (3.54).

From the definitions connected with Eq. (3.11) we have $k_c^2 = k_x^2 + k_y^2$ so that

$$(k_c)_{mn} = \sqrt{k_x^2 + k_y^2} = \sqrt{\left(\frac{m\pi}{a}\right)^2 + \left(\frac{n\pi}{b}\right)^2} \tag{3.55}$$

for *both* TM and TE modes. The cutoff wavelength and frequency may then be written

for TM and TE modes

$$(\lambda_c)_{mn} = \frac{2\pi}{k_c}\frac{2}{\sqrt{(m/a)^2 + (n/b)^2}}$$

$$= \frac{2ab}{\sqrt{(mb)^2 + (na)^2}} \tag{3.56}$$

$$(f_c)_{mn} = \frac{v}{(\lambda_c)_{mn}} = \frac{k_c}{2\pi\sqrt{\mu\epsilon}}$$

$$= \frac{1}{2\sqrt{\mu\epsilon}}\sqrt{\left(\frac{m}{a}\right)^2 + \left(\frac{n}{b}\right)^2} \tag{3.57}$$

where v is the velocity of the wave in an unbounded medium of dielectric constant ϵ and permeability μ. There are, therefore, doubly infinite sets of possible modes of each type corresponding to all the combinations of integers m and n. A TM mode with m *half-wave variations in the x direction and n half-wave variations in the y direction* is called a TM$_{mn}$ mode. Similar notation is used for the TE_{mn} modes. We note that for the TE_{mn} modes, either m or n (but not both) could be zero according to Eqs. (3.52), (3.30), and (3.31). However, for the TM_{mn} modes, neither m or n could be zero or otherwise the fields vanish completely according to Eqs. (3.51), (3.32), and (3.33). A transverse electromagnetic, or TEM, mode cannot propagate in a hollow waveguide, however, since by definition, $E_z = H_z = 0$ for such a mode.

Thus, according to Eq. (3.56), the lowest TE mode has the cutoff wavelength

$$(\lambda_c)_{TE_{10}} = 2a \qquad (3.58)$$

and the lowest TM mode has the cutoff wavelength

$$(\lambda_c)_{TM_{11}} = \frac{2ab}{\sqrt{a^2 + b^2}} \qquad (3.59)$$

Note that $(\lambda_c)_{TM_{11}} < (\lambda_c)_{TE_{10}}$ so that TE_{10} represents the dominant or lowest-order mode of a rectangular guide.

For purposes of computation, Eq. (3.57) may be put into the form

$$(f_c)_{mn} = \frac{1}{2a\sqrt{\mu\epsilon}} \sqrt{m^2 + \left[n\left(\frac{a}{b}\right)\right]^2} \qquad (3.60)$$

For TE_{10}, $(f_c)_{10} = 1/2a\sqrt{\mu\epsilon}$ and for TE_{20}, $(f_c)_{20} = 1/a\sqrt{\mu\epsilon}$ so that $(f_c)_{20}/(f_c)_{10} = 2$. Therefore, the maximum operating frequency range for a rectangular waveguide is

$$\frac{2(f_c)_{10} - (f_c)_{10}}{\frac{1}{2}[2(f_c)_{10} + (f_c)_{10}]} \times 100 = 66.7\%$$

which is the range if the waveguide were to propagate the dominant TE_{10} mode only. If $n \neq 0$, then $(f_c)_{mn}$ is a function of a/b.

We note from Fig. 3.2 that to obtain the maximum operating range for the TE_{10} mode, a/b should be about 2, as is the case in practice. Note also from Fig. 3.2 that in this case $(f_c)_{TE_{01}} = (f_c)_{TE_{20}} < (f_c)_{TM_{11}}$ or $(f_c)_{TE_{11}}$.

For $a/b = 2$, Eq. (3.60) becomes

$$(f_c)_{mn} = \frac{1}{2a\sqrt{\mu\epsilon}} \sqrt{m^2 + (2n)^2} \qquad (3.61)$$

Using Eq. (3.61), we can plot $(f_c)_{mn}$ of different modes on the frequency scale, as shown in Fig. 3.3. Figure 3.4 shows the field distribution of a few lower-order

Figure 3.3 Line spectrum of a rectangular waveguide. $a/b = 2$.

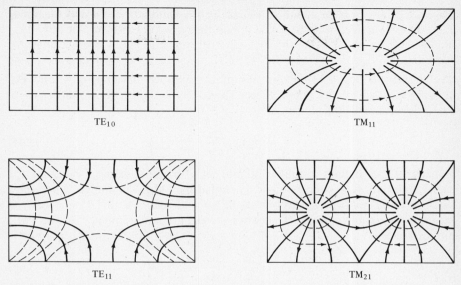

TE_{10} $\qquad\qquad$ TM_{11}

TE_{11} $\qquad\qquad$ TM_{21}

Figure 3.4 Field distribution of lower-order modes in a rectangular waveguide.

rectangular waveguide modes plotted with the help of Eqs. (3.51) and (3.52) and (3.30)–(3.33).

If $f > (f_c)_{mn}$, then those TE_{mn} or TM_{mn} modes, if excited, would propagate. Thus, a waveguide acts like a multiple cutoff filter. Since, for propagation, $f > (f_c)_{TE_{10}}$, it follows that $\lambda < (\lambda_c)_{TE_{10}} = 2a$ or $a > \lambda/2$ if the wave were to be above cutoff. But $(f_c)_{TE_{20}} = 2(f_c)_{TE_{10}}$, so $f < 2(f_c)_{TE_{10}}$ or $\lambda > (\lambda_c)_{TE_{10}}/2 = a$ if the TE_{20} mode is below cutoff. Therefore, to ensure single-mode propagation, we must have $\lambda/2 < a < \lambda$. Thus, *the cross-sectional dimensions of waveguides are usually on the order of a wavelength at microwave frequencies.*

3.4.2 Cylindrical Waveguides

We now turn to the study of cylindrical waveguides. In this case we have from Eq. (3.20) that

for TM modes
$$E_z = AJ_n(k_c r) \begin{cases} \cos n\theta \\ \sin n\theta \end{cases} \tag{3.62}$$

The Neumann function N_n is not included because it is infinite at $r = 0$ [see Fig. 3.1(b)], a point contained in the waveguide. Whether $\cos n\theta$ or $\sin n\theta$ is used is a matter of choice, for one could be obtained from the other by a 90° rotation. Likewise, we have from Eq. (3.20) that

for TE modes,
$$H_z = BJ_n(k_c r) \begin{cases} \cos n\theta \\ \sin n\theta \end{cases} \tag{3.63}$$

Again, N_n has not been included because it is infinite at $r = 0$. For TM modes $E_z = 0$ at $r = a$ according to Eq. (3.45) so that from Eq. (3.62)

$$J_n(k_c a) = 0 \tag{3.64}$$

The Bessel function $J_n(x)$ is zero for infinite number of values of x, and Eq. (3.64) is satisfied by any one of these. Thus, if p_{nl} is the lth root of $J_n(k_c a) = 0$, Eq. (3.64) is satisfied if

$$(k_c)_{nl} = \frac{p_{nl}}{a} \tag{3.65}$$

Equation (3.65) defines a doubly infinite set of possible values for k_c, one for each combination of integers n and l denoted by TM_{nl}. The integer n *describes the number of half-wavelength variations circumferentially, and l describes the number of variations radially.* The cutoff wavelength and frequency for the TM_{nl} modes are

$$(\lambda_c)_{\text{TM}_{nl}} = \frac{2\pi}{k_c} = \frac{2\pi a}{p_{nl}} \tag{3.66}$$

$$(f_c)_{\text{TM}_{nl}} = \frac{k_c}{2\pi\sqrt{\mu\epsilon}} = \frac{p_{nl}}{2\pi a\sqrt{\mu\epsilon}} \tag{3.67}$$

The lowest value of p_{nl} is the first root of the zero-order Bessel function, denoted p_{01} and equal to 2.405, so TM_{01} is the lowest TM_{nl} mode. From Eq. (3.66) we find that $\lambda_c = 2.61a$. Note that this wavelength is measured at the velocity of light ($= 1/\sqrt{\mu\epsilon}$) in an unbounded dielectric with dielectric constant ϵ and permeability μ.

For TE modes the required boundary condition is that normal derivatives of H_z be zero at $r = a$, according to Eq. (3.50). It then follows from Eq. (3.63) that

for TE modes $\qquad\qquad\qquad J_n'(k_c a) = 0 \tag{3.68}$

If p_{nl}' is the lth root of $J_n'(k_c a) = 0$, Eq. (3.68) is satisfied when

$$(k_c)_{nl} = \frac{p_{nl}'}{a} \tag{3.69}$$

Equation (3.69) again defines a doubly infinite set of possible TE_{nl} modes. Their cutoff wavelengths and frequencies are

$$(\lambda_c)_{\text{TM}_{nl}} = \frac{2\pi}{p_{nl}'} a \tag{3.70}$$

$$(f_c)_{\text{TE}_{nl}} = \frac{p_{nl}'}{2\pi a\sqrt{\mu\epsilon}} \tag{3.71}$$

The lowest value of p_{nl}' is not p_{01}' but rather $p_{11}' = 1.84$, so the TE_{11} mode has the lowest cutoff frequency of all transverse electric modes in a given diameter of guide. From Eq. (3.70) we find that this corresponds to a cutoff wavelength of $3.41a$. Thus, the TE_{11} mode has the lowest cutoff frequency in a circular guide.

The transverse components of **E** or **H** can be derived from Eqs. (3.30) and (3.31) for TE modes and from Eqs. (3.32) and (3.33) for TM modes. The field distribution of several lower-order modes in cylindrical waveguides is plotted in Fig. 3.5 with the help of Eqs. (3.62) and (3.63) and (3.30)–(3.33).

From the foregoing discussion we see that

$$(f_c)_{TM_{01}}/(f_c)_{TE_{11}} = 2.405/1.84 = 1.308$$

or a maximum single-mode operating range of

$$\frac{3.41 - 2.61}{\frac{1}{2}(3.41 + 2.61)} \times 100 = 26.6\%$$

for a circular waveguide propagating the TE_{11} mode. This compares with 66.7% for a rectangular waveguide propagating the TE_{10} mode, as calculated earlier. This difference in operating range coupled with the fact that the shape of the

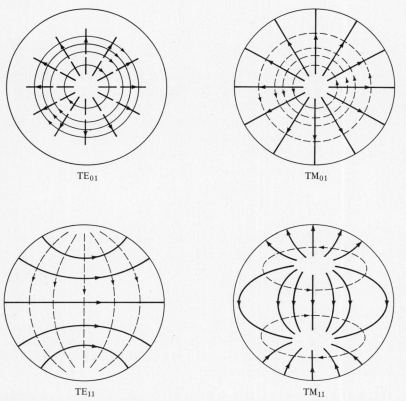

$$TE_{01} \qquad\qquad TM_{01}$$

$$TE_{11} \qquad\qquad TM_{11}$$

Figure 3.5 Field distribution of lower-order modes in a cylindrical waveguide.

rectangular guide fixes the polarization of the wave has resulted in a wider use of rectangular guides in microwave test equipment. When a circular waveguide has to be used, as in some applications, care must be taken to ensure that the polarization does not change due to the presence of obstacles, twists, or bends.

3.4.3 Coaxial Lines

We now turn to a brief discussion of coaxial lines, which frequently may be used at the lower end of the microwave frequency spectrum. Since the center conductor carries current, transverse $\mathbf{H}$ lines may exist by forming concentric circles enclosing the current. Transverse $\mathbf{E}$ lines are then radial—perpendicular to the $\mathbf{H}$ circles—starting and ending on charges on the inner and outer conductors. This coaxial line mode is designated as the transverse electromagnetic, or TEM, mode, thus signifying the absence of longitudinal field components.

Let us begin our discussion of the possible existence of a TEM mode in a coaxial guide by rewriting Eqs. (3.28) and (3.29) for $\mathbf{E}_t$ and $\mathbf{H}_t$ in terms of E_z and H_z:

$$\mathbf{E}_t = -\frac{\gamma}{k_c^2} \nabla_t E_z + \frac{\gamma_0 Z_0}{k_c^2} (\hat{\mathbf{z}} \times \nabla_t H_z) \tag{3.72}$$

$$\mathbf{H}_t = -\frac{\gamma}{k_c^2} \nabla_t H_z - \frac{\gamma_0}{k_c^2 Z_0} (\hat{\mathbf{z}} \times \nabla_t E_z) \tag{3.73}$$

We note from these equations that $\mathbf{E}_t$ and $\mathbf{H}_t$ are zero if E_z and H_z are both zero, indicating the nonexistence of a TEM mode, *unless* k_c is also zero rendering $\mathbf{E}_t$ and $\mathbf{H}_t$ indeterminate. Since $\omega_c = k_c/\sqrt{\mu\epsilon}$, it follows that the cutoff frequency is also zero, i.e., all frequencies, including dc, can propagate in this case. Inasmuch as the TEM mode must satisfy the same boundary conditions at all frequencies, we can expect the electric and magnetic fields to be of the following form:

$$\mathbf{E}_t = \mathbf{E}_{t0} e^{j\omega t - \gamma z} \tag{3.74}$$

$$\mathbf{H}_t = \mathbf{H}_{t0} e^{j\omega t - \gamma z} \tag{3.75}$$

where $\mathbf{E}_{t0}$ and $\mathbf{H}_{t0}$ are static solutions to Maxwell's equations.

For the dc case Maxwell's equations (2.1) and (2.2) become

$$\nabla \times \mathbf{E} = 0 \tag{3.76}$$

$$\nabla \times \mathbf{H} = \mathbf{i} \tag{3.77}$$

Since we have already noted that a TEM mode for which $k_c = 0$ cannot exist in a hollow guide, we might think of introducing a current-carrying coaxial conductor into a circular waveguide, as shown in Fig. 3.6. It then follows from Eq. (3.77) and Gauss's theorem that

$$\int \mathbf{H} \cdot d\mathbf{l} = I_0 \tag{3.78}$$

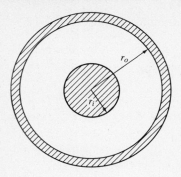

Figure 3.6   Coaxial line.

From symmetry considerations, we expect $\mathbf{H} = \hat{\boldsymbol{\theta}} H_\theta$ with H_θ a function of r but not of θ. Integrating around a circular path of radius r, we find from Eq. (3.78) that

$$\mathbf{H}_{t0} = \hat{\boldsymbol{\theta}} H_\theta = \hat{\boldsymbol{\theta}}\,\frac{I_0}{2\pi r} \tag{3.79}$$

Returning now to Eq. (3.76), we find that since $\nabla \times \mathbf{E} = \mathbf{0}$, $\mathbf{E} = -\nabla\phi$ where ϕ is a scalar potential. Since $\nabla \cdot \mathbf{D} = \rho = 0$ in this case, we have

$$\nabla^2\phi = \frac{1}{r}\frac{\partial}{\partial r}\left(r\frac{\partial\phi}{\partial r}\right) = 0 \tag{3.80}$$

This means that $r\,\partial\phi/\partial r = C$. Integrating this equation and evaluating the integration constant C, we find

$$\phi = \frac{V_0\ln(r/r_i)}{\ln(r_0/r_i)} \tag{3.81}$$

It further follows that

$$\mathbf{E}_{t0} = \hat{\mathbf{r}} E_r = -\hat{\mathbf{r}}\,\frac{\partial\phi}{\partial r} = -\hat{\mathbf{r}}\,\frac{V_0}{r\ln(r_0/r_i)} \tag{3.82}$$

Combining Eqs. (3.79) and (3.82), we find

$$\frac{E_r}{H_\theta} = \frac{2\pi V_0}{I_0\ln(r_0/r_i)} \tag{3.83}$$

To complete our analysis, we should also find the characteristic imped-ance Z_0 and propagation constant γ. First, from Eq. (3.5) we find, for $k_c = 0$,

$$\gamma = j\omega\sqrt{\mu\epsilon} \tag{3.84}$$

Then from Maxwell's equation (2.1), we find

$$\frac{E_r}{H_\theta} = \sqrt{\frac{\mu}{\epsilon}} = \eta \tag{3.85}$$

where $\eta = 377 \ \Omega$ is the free-space impedance. Combining Eqs. (3.83) and (3.85), we finally obtain

$$Z_0 = \frac{V_0}{I_0} = \frac{\eta}{2\pi} \ln\left(\frac{r_0}{r_i}\right) \tag{3.86}$$

where $\eta/2\pi = 60 \ \Omega$ for free space.

In addition to the TEM mode described above, TM_{np} and TE_{np} modes could also exist in a coaxial line. We would like to find the cutoff frequencies of the lowest TM and TE modes so that we may adjust the dimensions of the line to avoid them for a given range of operating frequencies.

From our general solution (3.20), we have

for TM modes
$$A_n J_n(k_c r_i) + B_n N_n(k_c r_i) = 0$$
$$A_n J_n(k_c r_0) + B_n N_n(k_c r_0) = 0 \tag{3.87}$$

or
$$\frac{N_n(k_c r_i)}{J_n(k_c r_i)} = \frac{N_n(k_c r_0)}{J_n(k_c r_0)} \tag{3.88}$$

since $E_z|_{r=r_i,r_0} = 0$ (see Fig. 3.6) according to Eq. (3.45).

For TE modes the derivative of H_z normal to the two conductors must be zero at r_i and r_0, according to Eq. (3.50). Thus,

for TE modes
$$C_n J'_n(k_c r_i) + D_n N'_n(k_c r_i) = 0$$
$$C_n J'_n(k_c r_0) + D_n N'_n(k_c r_0) = 0 \tag{3.89}$$

or
$$\frac{N'_n(k_c r_i)}{J'_n(k_c r_i)} = \frac{N'_n(k_c r_0)}{J'_n(k_c r_0)} \tag{3.90}$$

Solution of the transcendental equations (3.88) and (3.90) gives k_c as a function of r_i and r_0.

If the radius of curvature is large, we may imagine the conductors cut at plane A-A' of Fig. 3.7 and flattened into parallel planes. Recalling that for TM

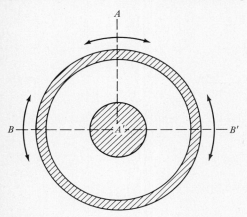

Figure 3.7   Other waveguide configurations evolved from the coaxial line.

modes $E_z = 0$ at $r = r_i, r_0$, we would expect that a TM_{op} mode will not propagate if

$$(r_0 - r_i) < \frac{p\lambda}{2} \tag{3.91}$$

where p is an integer. Thus, the cutoff wavelength is

$$\lambda_c = \frac{2}{p}(r_0 - r_i) \tag{3.92}$$

Actual solution of Eq. (3.88) indicates that Eq. (3.92) is correct near $r_0/r_i = 1$. Thus, the lowest-order TM mode is TM_{01}.[4]

We expect the cutoff for the lowest-order TE wave (TE_{11}) to occur when the average circumference of the coaxial line is about equal to a wavelength. Thus, if these modes were not to propagate,

$$2\pi \frac{r_0 + r_i}{2} < n\lambda \tag{3.93}$$

where n is an integer. The cutoff wavelength for the TE_{11} mode is accordingly

$$\lambda_c = \frac{2\pi}{n}\left(\frac{r_0 + r_i}{2}\right) \tag{3.94}$$

Calculation shows that Eq. (3.94) is valid to within 4% for r_0/r_i of up to 5. Since

$$\frac{2\pi}{n}\left(\frac{r_0 + r_i}{2}\right) > \frac{2}{p}(r_0 - r_i) \tag{3.95}$$

for $p = n = 1$ so that TE_{11}, aside from the TEM mode, is the lowest mode in a coaxial line. Electromagnetic field distributions for several lower-order modes are shown in Fig. 3.8.

3.4.4 Striplines and Microstrip Lines

A number of other guide configurations can be conceptually evolved from the coaxial line shown in Fig. 3.7. First, if the outer conductor of the coax is cut axially at position A and flattened downward as shown, we have the wire-over-ground configuration of Fig. 3.9(a). On the other hand, if the outer conductor is cut at plane B-B and flattened upward and downward as shown, the stripline configuration of Fig. 3.9(b) results provided the center conductor is imagined to be flattened into a strip and the entire region between top and bottom planes filled by a dielectric of relative dielectric constant ϵ_r. Finally, the microstrip line configuration of Fig. 3.9(c) results by essentially retaining only the lower half of the stripline shown in Fig. 3.9(b).

---

[4]For the TM case a plot of $\lambda_c/2(r_0 - r_i)$ vs r_0/r_i is given by S. Ramo, J. R. Whinnery, and T. Vanduzer, *Field and Waves in Communication and Electronics*, Wiley, New York, 1965, p. 447.

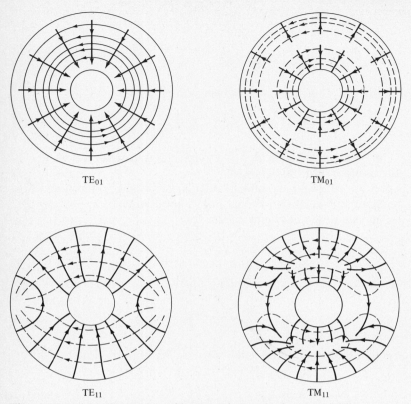

TE$_{01}$ TM$_{01}$

TE$_{11}$ TM$_{11}$

Figure 3.8 Field distribution of lower-order modes in a coaxial line.

The metal and plastic sandwich forming a stripline, as depicted in Fig. 3.9(b), can be produced at low cost by printed-circuit etching techniques, replacing heavy and costly ordinary waveguide components. This shielded stripline is free of radiation loss and is, therefore, suited for all types of circuits, including high-Q filters; the major part of the field is confined to a region within

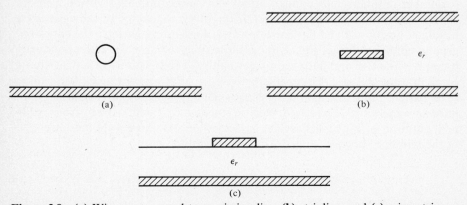

(a) (b) ϵ_r

ϵ_r

(c)

Figure 3.9 (a) Wire-over-ground transmission line, (b) stripline, and (c) microstrip.

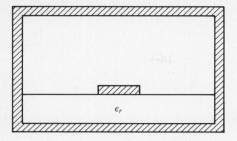

Figure 3.10 Microstrip line bounded by a shielding wall.

the vicinity of the strip and the structure supports a near-TEM mode. Figure 3.10 shows how a microstrip can be shielded.

An exact formula for the characteristic impedance Z_0 of a stripline with zero-strip thickness has been given,[5] but even a small thickness can affect the Z_0 substantially. For a line consisting of a circular conductor of diameter d centered between two parallel ground planes separated by a distance b, an approximate formula for Z_0 is[6]

$$Z_0 = \frac{60}{\sqrt{\epsilon_r}} \ln \frac{4b}{\pi d} \tag{3.96}$$

For a stripline with a center strip of width w and thickness t, d of Eq. (3.96) represents an equivalent diameter given by

$$d = 0.67w \left(0.8 + \frac{t}{w} \right) \tag{3.97}$$

which is applicable for $w/(b - t) < 0.35$ and $t/b < 0.25$. Equation (3.97) represents a linearization of the curve derived by Flammer for a long narrow antenna surrounded by a concentric cylinder having a large radius.[7]

The microstrip lines depicted in Fig. 3.9(c) are ideal for microwave integrated circuits because they have free surfaces on which to mount active and passive components. They are also extensively used to interconnect high-speed logic circuits in digital computers. Because of the radiative nature of their open structure, microstrip lines were initially abandoned by microwave engineers in favor of the shielded striplines. However, it was subsequently found that the use of thin high dielectric constant materials greatly reduces the radiative losses to render them useful in high-frequency applications.

Unlike the waveguides and coaxial lines discussed in Sections 3.4.1–3.4.3, the solution of the boundary value problem in connection with striplines is not straightforward. There are essentially two methods by which the characteristic impedance of a stripline can be found. In the field method the boundary value problem is solved with the aid of digital computers, yielding accurate results. In the second method approximate results are obtained by modifying known

[5]F. Oberttinger and W. Magnus, *Anwendung der Elliptischen Funktionen in Physik und Technik*, Springer-Verlag, Berlin, 1949, p. 63.

[6]S. B. Cohn, *IEEE Trans. Microwave Theory Tech.* **MTT-3**, 119 (1955).

[7]C. Flammer, "Equivalent Radii of Thin Cylindrical Antennas with Arbitrary Cross Sections," Stanford Research Institute Tech. Rep. (March 15, 1950).

expressions for input impedance of related structures. We shall discuss the second method first.

The characteristic impedance Z_0 of the wire-over-ground configuration depicted in Fig. 3.9(a) is

$$Z_0 = \frac{1}{2\pi} \sqrt{\frac{\mu_0}{\epsilon_0 \epsilon_r}} \ln \frac{4h}{d} = \frac{60}{\sqrt{\epsilon_r}} \ln \frac{4h}{d} \tag{3.98}$$

for $h \gg d$ and we have assumed that the surrounding medium is of relative dielectric ϵ_r. Equation (3.98) can easily be derived if we realize that Z_0 given by Eq. (3.98) should be equal to one-half the value of a two-wire line of spacing $2h$. This is because the potential distribution above the ground plane remains unchanged if the ground plane is replaced by the image of the wire at equal distance below ground. With the coaxial TEM mode expression for H_θ given by Eq. (3.79) and an expression similar to Eq. (3.82) for E_θ, superposition may be used to find the field distribution for a two-wire line. The characteristic impedance found is

$$Z_0 = \frac{1}{\pi} \sqrt{\frac{\mu_0}{\epsilon_0 \epsilon_r}} \ln \left(\frac{4h}{d} - 1 \right) \tag{3.99}$$

If $d \ll 4h$, this reduces to

$$Z_0 = \frac{1}{\pi} \sqrt{\frac{\mu_0}{\epsilon_0 \epsilon_r}} \ln \left(\frac{4h}{d} \right) \tag{3.100}$$

As mentioned previously, the characteristic impedance of a wire-over-ground line is just half this value, as quoted in Eq. (3.98). If effective or equivalent values of ϵ_r and d can be found for a microstrip line of strip width w and strip thickness t, an empirical (and very useful) expression similar in appearance to Eq. (3.100) can be formulated for the characteristic impedance of a microstrip line.[8]

For a plane wave propagating in an infinite medium of permeability μ and dielectric constant ϵ, the velocity of propagation is $1/\sqrt{\mu\epsilon}$, while the corresponding delay time per unit length is

$$T_d = \sqrt{\mu\epsilon} \tag{3.101}$$

Of course, owing to the fringing fields at the edge of the microstrip, the wave propagating along a microstrip line will not be exactly of the plane-wave type. This means that T_d will be a function of w as well as dependent on ϵ in a way more complicated than indicated in Eq. (3.101). Fortunately, it turns out that the dependence of T_d on w is quite weak, so that by inserting the measured values of T_d into Eq. (3.101), the equivalent dielectric constant ϵ_e to be substituted for ϵ in Eq. (3.98) can be found. The empirical expression for the effective relative dielectric constant ϵ_{re} so determined is

$$\epsilon_{re} = 0.475\epsilon_r + 0.67 \tag{3.102}$$

for several board materials such as fiberglass-epoxy and nylon phenolic of relative dielectric constant ϵ_r.

[8]H. R. Kaup, *IEEE Trans. Electron Comput.* **EC-16**, 185 (1967).

To use Eq. (3.98) for the characteristic impedance of a microstrip line, it remains necessary to relate the diameter d of the wire-over-ground configuration of Fig. 3.9(a) to the width w and thickness t of the microstrip line shown in Fig. 3.9(c). The empirical equation applicable for $0.1 \leq t/w \leq 0.8$ is the same as Eq. (3.97). Combining Eqs. (3.97), (3.98), and (3.102), we finally arrive at the expression for Z_0 of a microstrip:

$$Z_0 = \frac{87}{\sqrt{\epsilon_r + 1.41}} \ln \left(\frac{5.98h}{0.82w + t} \right) \tag{3.103}$$

By comparing the measured values of Z_0 with that computed from Eq. (3.103), Kaup found that Eq. (3.103) is applicable over a broad range of values of ϵ_r, h, w, and t.[9] We now proceed to discuss the analysis of microstrip lines by conformed mapping or computational means.

As early as 1952 Assadourian and Rimai developed a simplified theory for the microstrip transmission line using conformal transformation.[10] However, since they assumed that the strip is surrounded by a uniform medium, either air or dielectric, their analysis was not a rigorous one. In 1965 an accurate analysis was given by Wheeler.[11] He evaluated the transmission line properties of a microstrip line using conformal mapping of the dielectric boundary. All relations were approximated in terms of ordinary functions (no more advanced than exponential and hyperbolic) and some useful design curves were presented. However, whereas Assadourian and Rimai discussed the case of small strip thickness, Wheeler's analysis was restricted to strips of zero thickness.

An analytical program for calculating the field distribution about a microstrip transmission line bounded by a shielding wall (see Fig. 3.10) was used to calculate the impedance, velocity, and attenuation parameters.[12] In contrast to the "formula methods" cited above, exact solutions were obtained for TEM propagation using digital computation. A microstrip line, because it is partially filled, will not support a TEM mode, which means that a characteristic impedance of the line is not defined. However, when high-dielectric-constant substrate materials are used and the dimensions of the line and mode area are small compared to a half-wavelength, impedance and relative velocity as a function of impedance can be obtained to a good approximation. These mode area restrictions are necessary to limit cross talk between portions of the same circuit or propagation by other modes within the chamber.

Theoretical analysis of microwave propagation on microstrip lines with reference to the case of coupled pairs was carried out rigorously using a "dielectric Green's function" that expresses the discontinuity of fields at the dielectric-vacuum interface.[13] Data of this type are needed for the design of directional couplers, filters, and other components in microwave integrated circuits.

[9]Ibid., p. 185.

[10]F. Assadourian and E. Rimai, *Proc. Inst. Radio Eng.* **40**, 1651 (1952).

[11]H. A. Wheeler, *IEEE Trans. Microwave Theory Tech.* **MTT-13**, 172 (1965).

[12]H. E. Stinehelfer, *IEEE Trans. Microwave Theory Tech.* **MTT-16**, 439 (1968).

[13]T. G. Bryant and T. A. Weiss, *IEEE Trans. Microwave Theory Tech.* **MTT-16**, 1021 (1968).

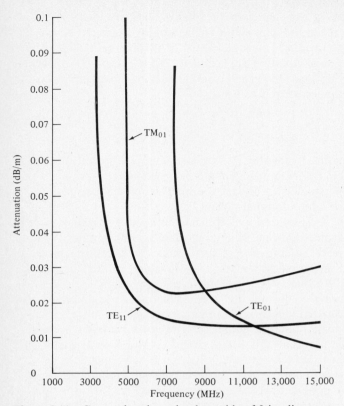

Figure 3.11   Copper loss in a circular guide of 2-in. diameter.

For a nonmagnetic dielectric substrate there are dielectric losses in the substrate and ohmic losses in the strip conductor and ground plane. Expressions have been derived for these losses in a microstrip line, taking into account the finite thickness of the strip.[14] It was found that in a microstrip line over a low-loss dielectric substrate, the predominant sources of loss at microwave frequencies are the imperfect conductors.

In addition to dielectric and ohmic losses, a microstrip line also has radiation losses that depend on substrate thickness and dielectric constant as well as on the stripline geometry.[15]

3.4.5 Dielectric Image Lines

Waveguides, coaxial lines, and striplines work well in the microwave frequency range, which extends roughly from 10^9 to 10^{11} Hz. For the millimeter wave

[14]R. A. Pucel, D. J. Masse, and C. P. Hartwig, *IEEE Trans. Microwave Theory Tech.* **MTT-16**, 342, 1064 (1968).

[15]L. Lewin, *Radiation from Discontinuities in Strip-line*, IEEE Monogr. No. 358E, February 1960.

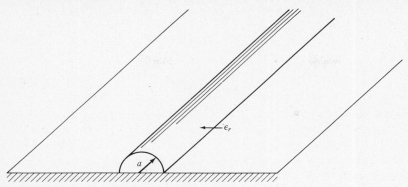

Figure 3.12 Dielectric image line.

band extending from 10^{11} to 10^{12} Hz, use of these types of transmission lines is not feasible because of their high attenuation and small component size.

One possibility for circumventing the high-loss problem at millimeter wave frequencies is to use the TE_{01} instead of the dominant TE_{11} mode in a circular guide. As can be seen from Fig. 3.11, unlike the other lower-order modes in rectangular and circular guides, the attenuation per unit length of the TE_{01} mode decreases continuously with frequency. In practice, however, in order to obtain the desired low attenuation, it is necessary to employ waveguides of sufficiently large diameter. Unfortunately, for a pipe of such diameter, many modes can propagate; e.g., at 55 GHz, 224 modes are above cutoff in a 2-in. pipe. Under these conditions, the TE_{01} mode is unstable against conversion to other modes of high attenuation. Mode conversion, particularly to the TE_{11} mode, which has the same propagation constant, occurs at slight bends in the guide or departures from circular cross section. Supression of mode conversion is possible but usually difficult and incomplete.

A possible low-loss millimeter waveguiding structure is the dielectric-metal waveguide or dielectric image line, shown in Fig. 3.12. We shall now discuss their characteristics in some detail. Unlike the stripline case, the dielectric image line problem can be solved exactly by analytical means.

We begin our discussion by solving the boundary value problem for a cylindrical dielectric waveguide. According to Schelkunoff,[16] the solution to wave equation (3.14) for E_z (and a similar equation for H_z) for propagation along a dielectric rod of radius a and dielectric constant ϵ is, for $r < a$ (inside the rod),

$$E_z = A J_n\left(\frac{pr}{d}\right) \cos n\phi \tag{3.104}$$

$$H_z = B J_n\left(\frac{pr}{d}\right) \sin n\phi \tag{3.105}$$

[16] S. A. Schelkunoff, *Electromagnetic Waves*, Van Nostrand, New York, 1943, p. 425.

with the factor $e^{j(\omega t - \beta z)}$ understood. Similarly, for $r > a$ (outside the rod)

$$E_z = CK_n\left(\frac{qr}{d}\right)\cos n\phi \tag{3.106}$$

$$H_z = DK_n\left(\frac{qr}{d}\right)\sin n\phi \tag{3.107}$$

Here J_n is the Bessel function, and K_n, apart from a constant factor, is the Hankel function of the second kind of imaginary argument. We note from Eqs. (3.104)–(3.107) that for $n \neq 0$ neither E_z nor H_z is zero, indicating the existence of hybrid modes. As in other geometries, the transverse components can be obtained from the longitudinal components given by Eqs. (3.104)–(3.107) by means of Eqs. (3.28) and (3.29). The parameters p and q are related to the phase constant β by the following relations:

$$\beta^2 = \omega^2\mu_1\epsilon_1 - \left(\frac{p}{a}\right)^2 \tag{3.108}$$

$$\beta^2 = \omega^2\mu_2\epsilon_2 + \left(\frac{q}{a}\right)^2 \tag{3.109}$$

where μ_1, ϵ_1 and μ_2, ϵ_2 are the permeability and dielectric constants of the rod and surrounding medium, respectively. For the case where the rod is non-ferromagnetic and the surrounding medium is free space, $\mu_1 = \mu_2 = \mu_0$ and $\epsilon_2 = \epsilon_0$. Because $\lambda_0\omega = 2\pi/\sqrt{\mu_0\epsilon_0}$, where λ_0 is the free-space wavelength, Eqs. (3.108) and (3.109) can be solved simultaneously to yield

$$\frac{2a}{\lambda_0} = \frac{1}{\pi}\left(\frac{p^2 + q^2}{\epsilon_r - 1}\right)^{1/2} \tag{3.110}$$

where $\epsilon_r = \epsilon_1/\epsilon_0$ is the relative dielectric constant of the rod. This equation shows that there is a definite relationship between p, q, and the diameter-to-wavelength ratio of the rod $2a/\lambda_0$.

Equations (3.108) and (3.109) or Eq. (3.110) is obtained directly from Maxwell's equations. A second relation between p and q is found from the boundary conditions, yielding the secular equation:

$$(\epsilon_r f + g)(f + g) - n^2\left(\frac{\epsilon_r}{p^2} + \frac{1}{q^2}\right)\left(\frac{1}{p^2} + \frac{1}{q^2}\right) = 0 \tag{3.111}$$

where the functions f and g are given by

$$f = \frac{J_n'(p)}{pJ_n(p)} \tag{3.112}$$

$$g = \frac{K_n'(q)}{qK_n(q)} \tag{3.113}$$

Thus, for given values of a, λ_0 (or ω), and ϵ_r, Eqs. (3.110) and (3.111) yield a pair of values p, q, defining a mode propagated along the rod.

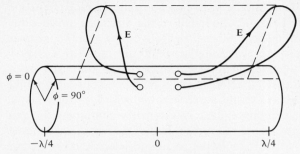

Figure 3.13 Typical field lines of "dipole" mode.

Since all functions involved are well known and tabulated, the simultaneous solution of Eqs. (3.110) and (3.111) by numerical methods presents no conceptual difficulties. Since the functions J_n and K_n are transcendental, there is an infinite number of solutions for each value of n. As Eqs. (3.104)–(3.107) show, a field of the nth-order mode depends on the azimuthal angle as $\cos(n\phi)$ or $\sin(n\phi)$. For $n = 0$, we have two rotationally symmetrical modes since Eq. (3.111) separates into two:

$$\epsilon_r f + g = 0 \tag{3.114}$$

$$f + g = 0 \tag{3.115}$$

The solutions of Eq. (3.114) give the transverse magnetic modes, while those of Eq. (3.115) give the circularly polarized transverse electric modes. It turns out, however, that these modes, with finite cutoff frequencies, are not the lowest-order modes; the lowest-order mode is the hybrid HE_{11}, or "dipole," mode with $n = 1$ and no radial node inside the rod. The dipole mode may be roughly described as a sinusoidal dielectric polarization perpendicular to the rod and traveling along it. The structure of the field is fairly complicated, and Fig. 3.13 shows a typical electric line of force, or rather its part outside of the rod. The magnetic field is similar but rotated 90° about the rod axis. There is no finite cutoff, so the rod can carry waves of arbitrary wavelength in this mode.

Elsasser[17] has analyzed the three modes (TE, TM with $n = 0$, and dipole with $n = 1$) in some detail. In particular, he computed the dielectric attenuation α_d along a rod of conductivity σ by using the fields obtained for the loss-free case given above. His result for the minimum-loss dipole mode, in decibels per meter, is

$$\alpha_d = 27.3 \left(\frac{\epsilon \phi_d}{\lambda_0} \right) R \tag{3.116}$$

where $\phi_d = \sigma/\epsilon_1 \omega$ is the power factor of the dielectric rod and R is a complicated function of the rod radius a and relative dielectric constant ϵ_r. For sufficiently small values of $2a/\lambda_0$, not only are the TE and TM ($n = 0$) modes below

[17]W. A. Elsasser, *J. Appl. Phys.* **20**, 1193 (1949).

cutoff, but the attenuation of the "dipole" mode itself is quite small. Under these conditions, the dipole mode is loosely bounded to the dielectric, or, in other words, the "mode area" is rather large. Since the effective cross section occupied by the wave is measured in square wavelengths, significant mode area and attendant large reductions in attenuation are only feasible in the millimeter band. Chandler[18] has investigated the characteristics of a dielectric transmission line at $\lambda_0 = 1.25$ cm and found that his results agree very well with Elsasser's theory.

Since the dipole mode exhibits a plane of symmetry containing the axis of the rod, an image plane may be used to replace half of the rod and surrounding space. Such an image system reduces the required dielectric cross section by one-half and also largely eliminates support and shielding problems connected with dielectric rod transmission lines. The resulting dielectric image line provides structural convenience and also has very low loss provided. As in the dielectric rod case, the wave is allowed to occupy a cross section many wavelengths square, which is readily achieved in the millimeter wave band. Indeed, straight sections of these lines have lower losses and are easier to fabricate than corresponding waveguides.

Attenuation of the image system with a perfectly conducting image plane is also given by Eq. (3.116). King and Schlesinger[19] have computed α_c, contribution of a lossy image surface to the line loss. Their result, in decibels per meter, is

$$\alpha_c = 69.5 \left(\frac{R_s}{\eta \lambda_0} \right) R' \tag{3.117}$$

where $R_s = \sqrt{\omega\mu/2\sigma}$ is the surface resistance of the conductor, $\eta = \sqrt{\mu_0/\epsilon_0}$ is the free-space impedance, and R' is a complicated function of a and ϵ_r. It was found that the conductor loss α_c is generally less than the dielectric loss α_d. In addition to these two types of losses, there is also radiation loss at bends, twists, and obstacles.[20] However, low attenuation and freedom from higher-order modes are the principal advantages of dielectric image lines operated in the HE_{11} dipole mode. In a further study Schlesinger and King[21] showed that there is excellent agreement between theory and experiment on a polystyrene image line. They also discussed results for an asymmetric line, the case of a dielectric bonding medium partially submerged in an image surface.

Attenuation of several dielectric image lines has been calculated for the 24–100-GHz range and checked with experiments performed at 35 and 70 GHz.[22] To obtain low attenuation, in addition to the requirement of low-loss dielectric and small cross section, low values of dielectric constant are also desirable. Several new foam plastics used were found suitable.

[18]C. H. Chandler, *J. Appl. Phys.* **20**, 1188 (1949).

[19]D. D. King and S. P. Schlesinger, *IEEE Trans. Microwave Theory Tech.* **MTT-5**, 31 (1959).

[20]Ibid., p. 31.

[21]S. P. Schlesinger and D. D. King, *IEEE Trans. Microwave Theory Tech.* **MTT-6**, 291 (1958).

[22]J. C. Wiltse, *IEEE Trans. Microwave Theory Tech.* **MTT-7**, 65 (1959).

3.5 IMPEDANCE, REFLECTION COEFFICIENT, AND SWR

In Section 3.4 we studied the field distribution in a waveguide in some detail. Here we wish to ascertain the impedance, reflection coefficient, and standing wave ratio (SWR) at a given cross-sectional plane of the waveguide when it is terminated in an arbitrary impedance or contains some form of discontinuity such as irises or junctions between waveguides of different cross sections. These problems have one feature in common: The obstacles set up a reflected wave traveling in the negative z direction. These reflected waves modify the field distribution both preceding and beyond the obstacles, thereby setting up standing waves.

If we change the direction of propagation from $+z$ to $-z$, the sign of γ changes because the forward and reflected waves vary, respectively, as $e^{j\omega t - \gamma z}$ and $e^{j\omega t + \gamma z}$. Therefore, the signs of Z_{TE} and Z_{TM} also change according to Eqs. (3.35) and (3.38). From Eqs. (3.31), (3.33), (3.36), and (3.37) we find that this brings about a change in sign in the relationship between the transverse components of $\mathbf{E}$ and $\mathbf{H}$ through Eqs. (3.36) and (3.37), in the relationship between $\mathbf{H}_t$ and H_z through Eq. (3.31), and in the relationship between $\mathbf{E}_t$ and E_z through Eq. (3.33). Letting the vector amplitude of the field components traveling in the $+z$ and $-z$ directions be designated by the subscripts $+$ and $-$, respectively, we have for the total field $\mathbf{E}$ and $\mathbf{H}$

$$\mathbf{E} = (\mathbf{E}_{t+}e^{j\omega t - \gamma z} + \mathbf{E}_{t-}e^{j\omega t + \gamma z}) + \hat{\mathbf{z}}(E_{z+}e^{j\omega t - \gamma z} - E_{z-}e^{j\omega t + \gamma z}) \quad (3.118)$$

$$\mathbf{H} = (\mathbf{H}_{t+}e^{j\omega t - \gamma z} - \mathbf{H}_{t-}e^{j\omega t + \gamma z}) + \hat{\mathbf{z}}(H_{z+}e^{j\omega t - \gamma z} + H_{z-}e^{j\omega t + \gamma z}) \quad (3.119)$$

If more than one mode exists in the waveguide, the right-hand side of Eqs. (3.118) and (3.119) should be summed over all modes. A check on the relative signs for the various terms in Eqs. (3.118) and (3.119) would show that they indeed satisfy the sign requirements enumerated above.

The question as to what is meant by the characteristic impedance of a waveguide now arises. Unlike the ordinary two-wire transmission line or the coaxial line propagating the TEM mode, the waveguide modes have transverse as well as longitudinal components of fields. Whereas an unambiguous impedance equal to the ratio of the transverse electric field to the transverse magnetic field, or the ratio of the voltage between conductors to the current in the conductor, can be defined for the TEM case, any definition of a waveguide "characteristic" impedance is complicated by the presence of the longitudinal field component of Eqs. (3.118) and (3.119). The matter is further complicated in that the field components are not constant over the cross section of the waveguide. For example, according to Eqs. (3.30) and (3.52), we have for the TE_{10} mode

$$\mathbf{E}_t = \hat{\mathbf{y}}\frac{\gamma_0 Z_0}{k_c^2}\frac{\partial H_z}{\partial x} = -\hat{\mathbf{y}}B\frac{\gamma_0 Z_0}{k_x}\sin k_x x \quad (3.120)$$

recalling that $k_c^2 = k_x^2 + k_y^2 = k_x^2$ in this case. It follows that the voltage across the top and bottom plates of the waveguide, being equal to $-\int \mathbf{E} \cdot d\mathbf{l}$, is not a constant but rather a function of x; it is maximum at the center and zero at the side walls. Similar remarks may be made regarding the magnetic field and, consequently, the current distribution.

A number of definitions for the characteristic impedance of a waveguide, each useful for certain purposes, can be proposed. Three of these, based on the maximum voltages V, total longitudinal current I, and power P, are given by Schelkunoff for the TE_{10} mode in a rectangular guide[23]

$$Z_{V,I} = \frac{V}{I} = \frac{\pi}{2} \frac{bE_y}{aH_x} \tag{3.121}$$

$$Z_{P,V} = \frac{V^2}{2P} = 2 \frac{bE_y}{aH_x} \tag{3.122}$$

$$Z_{P,I} = \frac{2P}{I^2} = \frac{\pi^2 bE_y}{8aH_x} \tag{3.123}$$

It is interesting to note that the three impedances defined on a voltage-current, power-voltage, and power-current basis are all proportional to E_y/H_x or the wave impedance. This is not particularly surprising, since the power flow down the waveguide is proportional to the z-directed Poynting vector or power density $\mathbf{P} = \mathbf{E}_t \times \mathbf{H}_t = \hat{z}E_tH_t$. Thus, it follows that $Z_{V,I}$, $Z_{P,V}$, and $Z_{P,I}$ differ only by differences in multiplicative constants.

Inasmuch as we are generally interested in the impedance at a particular plane of the waveguide relative to the characteristic impedance of the guide, however it may be defined, we can concentrate on the wave impedance or the ratio between the transverse components of the field. So long as a particular form of impedance is chosen, the arbitrary constants are divided out in the normalization, that is, in the ratio of impedance to characteristic impedance.

However, if two different waveguides are joined, the way in which the dimensions enter into an expression for the characteristic impedance must be considered. Even in this case, the numerical constants of Eqs. (3.121)–(3.123) involved need not be specified, as they will cancel in the expression for the impedance of one waveguide relative to the other as long as they are defined on the same basis. For example, if two waveguides whose cross-sectional dimensions are, respectively, a_1, b_1 and a_2, b_2, filled by dielectrics of constants μ_1, ϵ_1 and μ_2, ϵ_2 with corresponding free-space wavelengths λ_1 and λ_2, we find from Eq. (3.121), (3.122), or (3.123) that[24]

$$\frac{Z_1}{Z_2} = \frac{b_1 a_2}{b_2 a_1} \sqrt{\frac{\mu_1 \epsilon_2}{\mu_2 \epsilon_1}} \sqrt{\frac{1 - (\lambda_2/2a_2)^2}{1 - (\lambda_1/2a_1)^2}} \tag{3.124}$$

[23]S. A. Schelkunoff, *Electromagnetic Waves*, p. 319.

[24]G. L. Ragan, *Microwave Transmission Circuits*, MIT Radiation Laboratory Series, No. 9, McGraw-Hill, New York, 1948, p. 53.

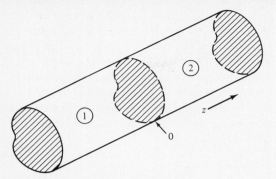

Figure 3.14   General waveguide.

where Z_1 and Z_2 are the characteristic impedances of waveguides 1 and 2, respectively, defined on the same basis. Actually Eq. (3.124) is applicable only if $a_1 \simeq a_2$ and $b_1 \simeq b_2$. For large changes in dimensions, it turns out that the real part of Z_1/Z_2 is still given by Eq. (3.124) but that a shunt susceptance appears in the equivalent circuit representing the excitation of nonpropagating modes at the discontinuity. By a quasistatic analysis it can be shown that this introduction of a shunt susceptance at a step discontinuity in a transmission line is characteristic of the behavior of sudden changes in line or guide dimensions.[25] In order to solve this problem exactly, one must treat it as a rather complicated boundary value problem in electromagnetic theory.[26]

In accordance with the above discussion, we shall restrict ourselves to the discussion of only the transverse components $\mathbf{E}_t$ and $\mathbf{H}_t$ in the calculation of the reflection coefficient, SWR, and so on. From Eqs. (3.118) and (3.119) we thus have

$$\mathbf{E}_t = \mathbf{E}_{t+}e^{j\omega t - \gamma z} + \mathbf{E}_{t-}e^{j\omega t + \gamma z} \tag{3.125}$$

$$\mathbf{H}_t = \mathbf{H}_{t+}e^{j\omega t - \gamma z} + \mathbf{H}_{t-}e^{j\omega t + \gamma z} \tag{3.126}$$

For the general case of irises and obstacles it is difficult to proceed further from Eqs. (3.125) and (3.126) for the relationship between $\mathbf{E}_{t+}$ and $\mathbf{E}_{t-}$, etc., may not easily be found. However, one type of reflection is particularly simple—the case in which the properties of the material filling the guide change discontinuously at $z = 0$, as shown in Fig. 3.14. In this case $\mathbf{E}_{t+} = \mathbf{E}_{t-}$, $\mathbf{E}_{t+} = Z_1\mathbf{H}_{t+}$, and $\mathbf{E}_{t-} = -Z_1\mathbf{H}_{t-}$ where Z_1 is the wave impedance of waveguide section 1. With the time dependence omitted, Eqs. (3.125) and (3.126) become

$$E_t = E_{t+}e^{-\gamma z} + E_{t-}e^{\gamma z} \tag{3.127}$$

$$H_t = \frac{1}{Z_1}(E_{t+}e^{-\gamma z} - E_{t-}e^{\gamma z}) \tag{3.128}$$

[25]S. Ramo, J. R. Whinnery, and T. Van Duzer, *Field and Waves*, p. 597.

[26]N. Marouwitz, *Waveguide Handbook*, MIT Radiation Laboratory Series, No. 10, McGraw-Hill, New York, 1948.

At $z = 0$, we have

$$\left.\frac{E_t}{H_t}\right|_{z=0} = Z_1 \frac{1 + E_{t-}/E_{t+}}{1 - E_{t-}/E_{t+}} \tag{3.129}$$

If waveguide 2 is terminated in its characteristic impedance or is infinite in length, there will be no reflected wave in section 2. It follows that $(E_t/H_t)_{z=0}$ is equal to the characteristic impedance Z_2 of section 2. Setting the right-hand side of Eq. (3.129) equal to Z_2 and solving for the ratio E_{t-}/E_{t+}, we find

$$\rho = \frac{E_{t-}}{E_{t+}} = \frac{Z_2/Z_1 - 1}{Z_2/Z_1 + 1} \tag{3.130}$$

The reflection coefficient ρ is defined as the ratio of the magnitude of the reflected wave to that of the incident wave.

The impedance as seen at any plane of medium 1, say at $z = -l$, can be obtained from Eqs. (3.127) and (3.128) by setting z in these equations equal to $-l$. In this way, we find

$$\frac{Z_{-l}}{Z_1} = \frac{(Z_2/Z_1)\cosh \gamma_1 l - \sinh \gamma_1 l}{\cosh \gamma_1 l - (Z_2/Z_1)\sinh \gamma_1 l} \tag{3.131}$$

where we have used Eq. (3.130).

The standing wave ratio S is defined as the ratio of the maximum electric field along the line E_{max} to the minimum electrical field along the line E_{min}:

$$S = \frac{E_{max}}{E_{min}} \tag{3.132}$$

E_{max} occurs at a point along the transmission line where the incidence and reflected waves interfere constructively or add in phase:

$$E_{max} = |E_{t+}| + |E_{t-}| \tag{3.133}$$

Correspondingly, E_{min} occurs at a point along the transmission line where the incidence and reflected waves interfere destructively or add out of phase by 180°:

$$E_{min} = |E_{t+}| - |E_{t-}| \tag{3.134}$$

Combining Eqs. (3.132)–(3.134), we have

$$S = \frac{|E_{t+}| + |E_{t-1}|}{|E_{t+}| - |E_{t-}|} \tag{3.135}$$

With Eq. (3.130), Eq. (3.135) becomes

$$S = \frac{1 + |\rho|}{1 - |\rho|} \tag{3.136}$$

The foregoing discussion is for a single discontinuity at a plane of constant z. The case of more than one discontinuity can be solved by repeated application of Eq. (3.131). Unlike the more complicated case of irises and obstacles, these problems can be handled exactly; their treatment will give us a good physical model for the attack on other cases.

A word may be said about the special cases embodied in Eq. (3.131). First, if the transmission line or waveguide 1 is lossless, Eq. (3.131) becomes

$$\frac{Z_{-l}}{Z_1} = \frac{(Z_2/Z_1)\cos\beta_1 l + j\sin\beta_1 l}{\cos\beta_1 l + j(Z_2/Z_1)\sin\beta_1 l} \tag{3.137}$$

where β_1 is the phase constant. If, in addition, waveguide 2 is replaced by a load impedance Z_L equal to Z_2, then Z_{-l}, as measured at the beginning of the transmission line, is appropriately called *input impedance* Z_{in}. Eq. (3.137) in this case becomes

$$\frac{Z_{\text{in}}}{Z_1} = \frac{(Z_L/Z_1)\cos\beta_1 l + j\sin\beta_1 l}{\cos\beta_1 l + j(Z_L/Z_1)\sin\beta_1 l} \tag{3.138}$$

This is a well-known formula in transmission line theory.

To summarize, we have seen that although different characteristic impedances for a waveguide can be defined differing in multiplicative constants, they are all proportional to the wave impedance. Therefore, as long as normalized impedances (impedance to characteristic impedance) are used, we need only concern ourselves with wave impedances or the correspondingly transverse field components. We note in this regard, from Eqs. (3.130) and (3.131), that indeed only ratios of impedances Z_2/Z_1 and Z_{-l}/Z_1 appear. For a more rigorous treatment of this problem based on normal modes and electromagnet boundary conditions, the reader is referred elsewhere.[27]

3.6 RESONANT CAVITIES

3.6.1 Lumped Analogs

Before we proceed to the study of microwave resonant cavities, it is instructive to review briefly low-frequency lumped resonant circuits. In particular, we shall see that the concept of a complex frequency is a natural consequence of our calculation.

For a series-resonant circuit composed of a resistance R, an inductance L, and a capacitance C, the current i is related to the driving voltage V by the following differential equation:

$$L\frac{di}{dt} + Ri + \frac{1}{C}\int i\,dt = V \tag{3.139}$$

If, after the current is established, the impressed voltage is removed, i will decay in amplitude as a function of time owing to the presence of losses. Assuming

[27]J. C. Slater, *Microwave Electronics*, Van Nostrand, New York, 1980, p. 22.

$i \sim e^{j\omega t}$, we have from Eq. (3.139)

$$R + j\left(\omega L - \frac{1}{\omega C}\right) = 0 \tag{3.140}$$

If we let $1/LC = \omega_0^2$ and $\omega_0 L/R = Q$, Eq. (3.140) becomes

$$\frac{1}{Q} + j\left(\frac{\omega}{\omega_0} - \frac{\omega_0}{\omega}\right) = 0 \tag{3.141}$$

Solving for ω in Eq. (3.141), we obtain

$$\omega = \omega_0 \sqrt{1 - \left(\frac{1}{2Q}\right)^2} + j\frac{\omega_0}{2Q} \tag{3.142}$$

It is significant to note that we have found the frequency ω to be complex. The physical significance of this can best be ascertained by letting $\omega = \omega_1 + j\omega_2$. Using this expression in Eq. (3.142), we find

$$\omega_0 = \sqrt{\omega_1^2 + \omega_2^2} \tag{3.143}$$

and

$$Q = \sqrt{\left(\frac{\omega_1}{2\omega_2}\right)^2 + \frac{1}{4}} \tag{3.144}$$

Since $i \sim e^{j\omega t}$, it follows that in terms of ω_1 and ω_2

$$i \sim e^{-\omega_0 t/2Q} \exp\left[j\omega_0 \sqrt{1 - \left(\frac{1}{2Q}\right)^2}\, t\right]$$

It is seen that in contrast to the case of a purely real frequency, the case of a frequency with a real as well as an imaginary part gives rise to an exponential decay with time of the magnitude of the current.

For a parallel circuit we again obtain equations of the same form as Eqs. (3.141)–(3.144), providing we define Q as equal to $RC\omega_0$ rather than $\omega_0 L/R$ as for the series case. Thus, we see that the definitions for Q for series and parallel resonant circuits are the same when defined in terms of the real and imaginary parts of the frequency. For this reason, it is more convenient to define Q in terms of Eq. (3.144) for distributed circuits such as those encountered in microwave electronics and magnetics.

3.6.2 Rectangular Cavity

Consider again the electromagnetic field in a rectangular waveguide supporting the TE_{10} mode. From Eqs. (3.52)–(3.54) and (3.30)–(3.31) we have

$$H_z = B \cos \frac{\pi x}{a} \tag{3.145}$$

$$H_x = B \frac{\pi}{a} \frac{\gamma}{k_c^2} \sin \frac{\pi x}{a} \tag{3.146}$$

$$E_y = -\frac{\gamma_0 Z_0}{k_c^2}\left(\frac{\pi}{a}\right) B \sin \frac{\pi x}{a} \tag{3.147}$$

If the direction of propagation is reversed, from $+z$ to $-z$, γ and therefore H_x change sign. A rectangular cavity can be formed by closing a rectangular waveguide with two shorting plates placed at a distance d apart and perpendicular to the z axis, as shown in Fig. 3.15. In this case the oppositely traveling waves in the $+z$ and $-z$ directions combine to yield

$$H_z = (Be^{-\gamma z} + B'e^{\gamma z})\cos\frac{\pi x}{a} \tag{3.148}$$

$$H_x = (Be^{-\gamma z} - B'e^{\gamma z})\left(\frac{\pi}{a}\right)\frac{\gamma}{k_c^2}\sin\frac{\pi x}{a} \tag{3.149}$$

$$E_y = -(Be^{-\gamma z} + B'e^{\gamma z})\frac{\gamma_0 Z_0}{k_c^2}\left(\frac{\pi}{a}\right)\sin\frac{\pi x}{a} \tag{3.150}$$

Since E_y must be equal to zero at $z = 0$ according to Eq. (3.44), we find from Eq. (3.150) that $B' = -B$. Thus, Eqs. (3.148)–(3.150) become

$$H_z = -2B\cos\frac{\pi x}{a}\sinh\gamma z \tag{3.151}$$

$$H_x = 2B\left(\frac{a}{\pi}\right)\gamma\sin\frac{\pi x}{a}\cosh\gamma z \tag{3.152}$$

$$E_y = B\gamma_0 Z_0\left(\frac{a}{\pi}\right)\sin\frac{\pi x}{a}\sinh\gamma z \tag{3.153}$$

where we have noted that $k_c = \pi/a$ from Eq. (3.53).

At $z = d$, E_y must likewise be zero according to Eq. (3.44). Since none of the z-independent factors in Eq. (3.153) is in general zero at $z = d$, it follows that $\sinh\gamma d$ must be equal to zero. Letting $\gamma = \alpha + j\beta$ and using a trigonometric transformation, we have

$$\sinh\alpha d\cos\beta d + j\cosh\alpha d\sin\beta d = 0 \tag{3.154}$$

For Eq. (3.154) to hold, we see that α must be zero. In this case Eq. (3.154)

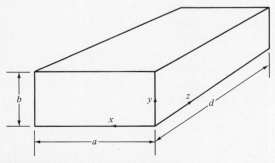

Figure 3.15 Rectangular cavity.

reduces to

$$\sin \beta d = 0 \quad \rightarrow \quad \beta = \frac{l\pi}{d} \tag{3.155}$$

where l is an integer. Equation (3.155) indicates that E_y must be zero at $z = 0$ and d so that the length of the rectangular cavity must be such that it corresponds to an integral number of half-wavelengths of the electric field.

Although in the calculation above we have implicitly assumed that the guidewalls are lossless, it is permissible for the dielectric filling the guide to have losses. This being the case, it is not at all clear how α in $\gamma = \alpha + j\beta$ could be zero with the complex frequency defined in Section 3.6.1. Letting $\omega = \omega_1 + j\omega_2$, we find from Eq. (3.6) that

$$\gamma = \sqrt{[k_c^2 - (\omega_1^2 - \omega_2^2)\epsilon\mu - \sigma\omega_2\mu] + j[\omega_1\mu(\sigma - 2\omega_2\epsilon)]} \tag{3.156}$$

If γ is to be imaginary, the expression in the first bracket must be negative while that in the second must be zero, i.e.,

$$[(\omega_1^2 - \omega_2^2)\mu\epsilon + \sigma\omega_2\mu] > k_c^2 \tag{3.157}$$

and

$$\omega_2 = \frac{\sigma}{2\epsilon} \tag{3.158}$$

Equation (3.157) is a generalization of the statement that the operating frequency must be above the cutoff value for propagation to occur in a waveguide containing a lossy dielectric. Equation (3.158), on the other hand, relates the imaginary part of the frequency to the conductivity and dielectric constant of the dielectric filling the cavity.

To recapitulate, since the fields are assumed to be proportional to $e^{j\omega t - \gamma z}$, the results obtained indicate that

$$E_y \propto e^{-\omega_2 t} e^{j\omega_1 t} \exp(j\{[\omega_1^2 - \omega_2^2)\mu\epsilon + \sigma\epsilon_2\mu] - k_c^2\}^{1/2} z)$$

and likewise for H_x and H_z. Since γ is purely imaginary, there is propagation without attenuation with z. In this connection, we may recall that the standing wave in the z direction is composed of two oppositely traveling waves. Thus, these waves travel along the $+z$ and $-z$ directions without an accompanying decrease in amplitude even though the dielectric filling the cavity is not lossless. In this case, as the energy is absorbed with the passage of time, the amplitude of these waves and that of their linear combination must decrease with time.

In this discussion we have implicitly assumed that the cavity is a completely enclosed structure. In practice, of course, this cannot be the case, as energy must be coupled into or away from the cavity. This coupling is usually achieved by having a small hole in one wall of the cavity. The coupling hole is small in order to prevent the bulk of the energy entering the cavity from reradiating and thereby lowering the total Q. Accordingly, the total Q of the cavity is given by the expression

$$\frac{1}{Q_{\text{total}}} = \frac{1}{Q_{\text{coupling}}} + \frac{1}{Q_{\text{dielectric}}} + \frac{1}{Q_{\text{wall}}} \tag{3.159}$$

where Q_{total} is usually known as the loaded Q, $Q_{coupling}$ is related to the energy reradiated out of the coupling hole (the external Q), and

$$\frac{1}{Q_{dielectric}} + \frac{1}{Q_{wall}} = \frac{1}{Q_{unloaded}} \tag{3.160}$$

The unloaded Q, in this sense, means the Q of the cavity "not loaded" by the presence of the coupling hole. This is important to remember because if the cavity were loaded by a sample whose properties are to be measured, there would be a change in the unloaded Q. In this case

$$\frac{1}{Q_{unloaded}} = \frac{1}{Q_{dielectric}} + \frac{1}{Q_{wall}} + \frac{1}{Q_{sample}} \tag{3.161}$$

We now turn to the calculation of the resonance frequency ω_1. Combining Eqs. (3.155)–(3.158), we find that

$$\omega_1 = \omega_0 \sqrt{1 - \left(\frac{1}{2Q}\right)^2} \tag{3.162}$$

where

$$\omega_0 = \frac{1}{\sqrt{\mu\epsilon}} \sqrt{\left(\frac{\pi}{d}\right)^2 + \left(\frac{l\pi}{d}\right)^2} \tag{3.163}$$

and

$$Q = \frac{\omega_0 \epsilon}{\sigma} \tag{3.164}$$

It is significant to note that Eq. (3.162) is of the same form as the real part of the corresponding Eq. (3.142) for the lumped circuit. Furthermore, it can be shown that the definitions for Q given by Eqs. (3.144) and (3.164) are equivalent. The same is true for ω_0 given by Eqs. (3.143) and (3.163). By induction, we can generalize Eq. (3.163) to read

$$\omega_0 = \frac{1}{\sqrt{\mu\epsilon}} \sqrt{\left(\frac{m\pi}{a}\right)^2 + \left(\frac{n\pi}{b}\right)^2 + \left(\frac{l\pi}{d}\right)^2} \tag{3.165}$$

where ω_0 is the resonance frequency of a rectangular cavity oscillating in the TE_{mnl} mode.

If the cavity walls have finite losses, the calculation of ω and Q is more complicated. However, if the losses of the walls are sufficiently small, Q_{wall} can be determined rather easily by an approximate calculation. In such a calculation the fields in the cavity are assumed to be the same as that of a lossless cavity. In particular, the component of H tangential to the perfectly conducting wall is equal to the surface current density in magnitude, according to Eq. (2.58). From the surface current densities at the various cavity walls and their conductivity, the ohmic losses can be calculated.

From Eqs. (3.142) and (3.143) we see that for $Q \gg \frac{1}{2}$, $\omega_0 \simeq \omega_1$. Accordingly, $Q \simeq \omega_1/2\omega_2 \simeq \omega_0/2\omega_2$ from Eq. (3.144). This definition of Q is equivalent to

$$Q = \frac{\omega_0 U}{W_L} \tag{3.166}$$

where U is the energy stored in the cavity and W_L is the average power loss. Equation (3.166) is easily derived from the solution i of Eq. (3.139) by noting that $U = \frac{1}{2}L|i|^2$ and $W_L = -dU/dt$. From knowledge of the field distribution in the cavity, U as well as W_L can be calculated. Q can then be determined via Eq. (3.166).

3.6.3 Other Types of Cavity

Besides the rectangular cavity, there are a number of cavities of other configurations. The coaxial cavity, cylindrical cavity, spherical cavity, reentrant cavity, and optical cavity are examples. Although the geometry of these cavities may be very different from the rectangular one, such concepts as resonance frequency and Q are still applicable. These cavities and methods of coupling to them will be studied as needed in other parts of this book.

3.7 VELOCITY, ENERGY, AND POWER FLOW

So far we have assumed that the fields are harmonic in time, that is, that **E**, **D**, **B**, and **H** are proportional to $\exp\{j\omega[t - (\beta/\omega)z]\}$ for the lossless case. Since $(\beta/\omega)z$ must have dimensions of time, it is clear that ω/β corresponds to some sort of velocity. It turns out that this velocity, known as the phase velocity v_p, is the velocity with which a plane of constant phase travels, for if $t - (\beta/\omega)z$ is independent of space and time, we have

$$t - \frac{\beta}{\omega} z = \text{const} \tag{3.167}$$

Differentiating Eq. (3.167), we find

$$v_p = \frac{dz}{dt} = \frac{\omega}{\beta} \tag{3.168}$$

Thus we see that the concept of a phase velocity applies only to fields that are periodic in space and consequently represent wave trains of infinite duration. A wave train of finite length, on the other hand, cannot be represented in a simple harmonic form and the term *phase velocity* loses its precise significance. To illustrate this point, let us consider the case of the superposition of two sinusoidal waves, each of unit amplitude, that differ very slightly in frequency and phase constant. In this case the resultant wave is represented by

$$\cos(\omega t - \beta z) + \cos[(\omega + \delta\omega)t - (\beta + \delta\beta)z]$$

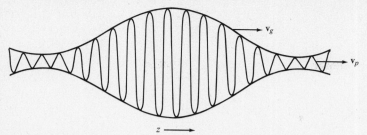

Figure 3.16 Group and phase velocities of a wave.

By a trigonometric transformation, this becomes

$$2 \cos \tfrac{1}{2}[(\delta\omega)t - (\delta\beta)z] \cos[(\omega + \tfrac{1}{2}\delta\omega)t - (\beta + \tfrac{1}{2}\delta\beta)z]$$

We see that the resultant field has a phase constant essentially equal to β and oscillates at a frequency that differs negligibly from ω. It then follows that the velocity of a plane of constant phase is still given essentially by v_p of Eq. (3.168). On the other hand, it is clear that the resultant amplitude, given by

$$2 \cos \tfrac{1}{2}[(\delta\omega)t - (\delta\beta)z]$$

varies slowly between the sum of the amplitudes of the component waves and zero. As a result of the interference of the two waves of slightly different frequency and phase constant, the field distribution in space and time has an envelope with a series of periodically repeated beats. In this case the velocity with which a plane of constant amplitude travels, called the group velocity v_g, is obtained from the relation

$$(\delta\omega)t - (\delta\beta)z = \text{const} \tag{3.169}$$

Differentiating, we find

$$v_g = \frac{dz}{dt} = \frac{\delta\omega}{\delta\beta} \tag{3.170}$$

The physical significance of these results is best illustrated by reference to Fig. 3.16. We see that the group velocity is the velocity of the envelope, while the phase velocity is the high-frequency wave within the envelope. For $v_p > v_g$, we would expect the high-frequency wave to slip through the envelope with the passage of time.

A pulse of any shape can be constructed by choosing the amplitude of the component waves as an appropriate function of ω or β and integrating by means of the Fourier integral. However, the concept of a group velocity applies only to a narrow-band spectrum in ω or β. As the interval $\delta\omega$ or $\delta\beta$ increases, the spread in the phase velocity of the component waves increases markedly provided the medium is dispersive, that is, that $\omega = f(\beta)$. In this case the packet (wave amplitude vs ω or β) is severely deformed and the concept of the group velocity becomes meaningless.[28]

[28]J. A. Stratton, *Electromagnetic Theory*, McGraw-Hill, New York, 1941, p. 331.

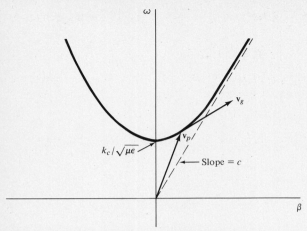

Figure 3.17 ω-β diagram for a waveguide.

To relate our discussion of v_p and v_g to the case of waveguides, let us further examine Eq. (3.5). Solving for $\omega\sqrt{\mu\epsilon}$ in that equation, we find

$$\omega\sqrt{\mu\epsilon} = \sqrt{k_c^2 + \beta^2} \tag{3.171}$$

where we have set $\gamma^2 = -\beta^2$ for a lossless medium. Referring to Fig. 3.17 or Eq. (3.171) we see that for $\beta \gg k_c$, that is, for a guide operating way above cutoff, $\omega/\beta \simeq 1/\sqrt{\mu\epsilon} = c$ where c is the velocity of light in the unbounded medium of dielectric constant ϵ and permeability μ. However, for smaller values of β, the relationship between ω and β is no longer linear, as can be seen from Fig. 3.17. In other words, $v_p = \omega/\beta$ is not a constant but rather a function of ω, for the waveguide is dispersive. On the other hand, differentiating Eq. (3.5), we have

$$v_g = \frac{d\omega}{d\beta} = \frac{\beta}{\mu\epsilon\omega} = \frac{1}{\mu\epsilon v_p} \tag{3.172}$$

so that $v_p v_g = c^2$, a characteristic of empty waveguides. Again, we find that v_g is a function of ω. In Fig. 3.17 we see that v_p is the slope of the line from the origin to a point on the curve, while v_g is the local slope at the same point.

Group velocity is often associated with the velocity of energy propagation. This identity is valid in many cases with normal dispersion ($dv_p/d\omega < 0$) such as the waveguides discussed here, but not usually for systems with anomalous dispersion ($dv_p/d\omega > 0$).[29] If v_E is designated as the velocity of energy flow, then

$$v_E = \frac{W_T}{U_{av}} \tag{3.173}$$

where W_T is the average power flow in a single mode and U_{av} is the average energy stored in a unit of guide length. In this case W_T is equal to the Poynting

---

[29]For a careful study of the wave front and signal velocities, see Ibid., p. 333.

vector **P** integrated over the cross section of the guide, and **P** is defined in terms of the fields as

$$\mathbf{P} = \mathbf{E} \times \mathbf{H} \tag{3.174}$$

For a lossless waveguide it is clear that **P** has only a z component, since no energy can be consumed by perfectly conducting walls. It then follows that

$$\mathbf{P} = \hat{z} E_t H_t$$

where E_t and H_t are transverse fields (see Section 3.5, the discussion of wave-guide characteristic impedance).

PROBLEMS

3.1 Figure 3.18 shows a metallic strip of conductivity σ and a thickness δ. If the electric field $E(\delta)$ at $x = \delta$ is equal to $E_0 e^{j\omega t}$, find the electric field in regions 2 and 3.

3.2 Show that the TEM mode cannot propagate in an empty waveguide.

3.3 Show two methods of excitation of the TE_{30} mode in a rectangular guide. What other modes would you expect to be excited by your methods?

3.4 Sketch the charge and current distribution on the guidewalls for the TE_{11} mode in a rectangular waveguide.

3.5 Using the method of images, show that a stripline with a conductor centered be-tween two parallel planes can be replaced by a series of conductors.

3.6 Derive Eq. (3.99).

3.7 Derive Eq. (3.111).

3.8 Sketch the electric and magnetic fields of the $n = 0$ TE and TM modes in a dielectric image line.

3.9 **(a)** Consider two waveguides connected by a junction. The reflection coefficient looking into the junction from waveguide 1 is ρ_1, and the reflection coefficient

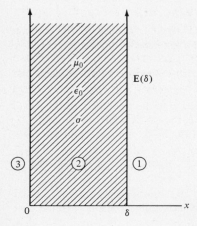

Figure 3.18 Metallic strip.

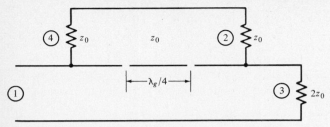

Figure 3.19 Directional coupler.

looking into waveguide 2 is ρ_2. If $\rho_1 \neq \rho_2$ and the junction contains no nonreciprocal elements, what can you deduce about the characteristic of the junction?

(b) For a unit of power entering waveguide 1, an amount of power δ emerges from the junction into waveguide 2. If a unit of power enters waveguide 2, what amount of power emerges from the junction into waveguide 1?

3.10 (a) Consider the directional coupler of Fig. 3.19. Explain how it works. Z_0 is the characteristic impedance of the guide, and λ_g is the guide wavelength.

(b) Given that one unit of power enters terminal 1, find the power dissipated in the terminating impedances at 2–4. Assume that the coupling coefficient between the guides is 3 db and the directivities are infinite.

(c) Find the voltage standing-wave ratio (VSWR) at terminal 1.

3.11 (a) Consider the case of two waveguides of characteristic impedance Z_{01} and Z_{03} connected by a third of characteristic impedance Z_{02} and length l. Find l as a function of frequency ω and the relationship between Z_{01}, Z_{02}, and Z_{03} given that all power entering guide 1 emerges from guide 2 into guide 3.

(b) Investigate the way in which the magnitude of the reflection coefficient at the 1-2 interface varies with frequency as it departs from the frequency for which the matching section is designed. Show that the section will be most broadbanded for the shortest matching section.

3.12 (a) Consider a lossless rectangular waveguide terminated by a short. Find the impedance, reflection coefficient, and SWR at place *A-A* located at a distance d in front of the short.

(b) If $d = l\lambda_g/2$, where l is an integer and λ_g is the TE_{10} mode guide wavelength, the shorted waveguide can be considered to be a cavity with a full-size coupling hole. Calculate the resonance frequency and Q of such a cavity.

3.13 (a) Expand the field distribution in the cavity described in Problem 3.12(b) in terms of its normal modes (TE_{mnl} and TM_{mnl}). Can a summation of these modes satisfy the boundary condition at plane *A-A*?

(b) If the answer to (a) is no, what is the correct field distribution inside the cavity?

3.14 (a) Consider a cubic rectangular resonator. Find Q of the resonator for the TE_{10l} modes. Assume that the surface resistivity of the cavity walls is R_s.

(b) Sketch Q of the resonator as a function of the mode number l and also as a function of frequency.

(c) Represent the cavity by a parallel *RLC* resonant circuit. Find the expressions for R, L, and C for the TE_{10l} modes.

3.15 Find the approximate resonance frequency of the foreshortened lossless coaxial cavity of Fig. 3.20 as a function of a, b, l, δ, etc. For simplicity, assume that

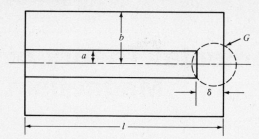

Figure 3.20 Foreshortened coaxial cavity.

$\delta \ll l$ and that the electric field is concentrated at the gap region G while the magnetic field is distributed throughout the rest of the cavity. Contrast this frequency with the lowest resonance frequency of the TEM mode in lossless cavity of length l.

3.16 An empty rectangular waveguide with dimensions $a = 4$ cm and $b = 2$ cm is excited at a frequency of 5 GHz in the fundamental mode. If the exciting source produces a maximum electric field strength of 10^4 V/m at the center of the entrance cross section, what is the power input to the guide?

3.17 (a) Show mathematically that no power can propagate in a lossless waveguide below cutoff even though the electromagnetic fields inside the guide are finite.

 (b) If a crystal detector is inserted into the cutoff waveguide of (a), will any power be extracted? Why?

Chapter 4

Fundamental Theory of Magnetism

In this chapter various types of magnetism are discussed. Diamagnetism, paramagnetism, ferromagnetism, antiferromagnetism, and ferrimagnetism are treated, and, in particular, the related susceptibilities are calculated in terms of the electronic orbital and spin contributions to the magnetization of the material. Exchange energy, domains, and domain walls in ferromagnets and ferrimagnets are also discussed. Although the discussion is mostly classical in nature, quantum-mechanical concepts are introduced wherever necessary. Every attempt has been made to present difficult concepts as simply and as clearly as possible without sacrifice of correctness and rigor.

4.1 INTRODUCTION

In Chapter 1 we briefly discussed the orbital and spin contributions of electrons to the magnetic properties of materials. We shall now elaborate on this subject. To begin with, in the presence of electric and magnetic fields electrons will in general execute some form of orbital motion. Consider specifically the simple case of the hydrogen atom. Considered as classical particles, both the proton and the electron rotate about a common center of mass with a common angular velocity. However, if the electron mass is replaced by the reduced mass, the proton can be considered as fixed in space while the electron executes orbital motion about the stationary proton. Thus, within the framework of classical mechanics we consider these particles to have a definite position and momentum at any given time. It further follows that the position and velocity of the particle can be definitively specified at any given time. As is well known, such assumptions are disavowed in quantum mechanics.

According to the uncertainty principle in quantum mechanics, if the position of a particle is definitely known, then we can have no knowledge whatsoever of its momentum. Thus, the classical concept of a particle circulating in a definite orbit with a definite velocity is untenable in quantum mechanics. Instead, the probability of a particle being in volume dv centered about some point $\mathbf{r}$ is given by the quantity $\psi\psi^* \, dv$ where ψ is known as the wave function of the particle. ψ can be calculated from the Schrödinger time-dependent wave equation[1]:

$$-\frac{\hbar^2}{2m}\nabla^2\psi + V\psi = j\hbar\frac{\partial\psi}{\partial t} \tag{4.1}$$

where V is the potential energy. If V is independent of time, we can easily show by the method of separation of variables that $\psi \sim e^{-jEt/\hbar}$ where E is the total energy. Accordingly, Schrödinger's equation takes its time-independent form:

$$\nabla^2\psi + \frac{2m}{\hbar^2}(E - V)\psi = 0 \tag{4.2}$$

Equations (4.1) and (4.2) can be thought of as summarizing experimental observations in atomic physics.

To solve these equations, it is necessary to specify the spatial dependence of V and to impose the boundary conditions for the continuity of ψ and its first derivative. For example, in the case of the hydrogen atom V is equal to $-e^2/4\pi\epsilon_0 r$. Inserting this expression for V into Eq. (4.2), we find the energy E to be

$$E = -\frac{m_r e^4}{2\hbar^2 n^2} \tag{4.3}$$

where m_r is the reduced mass[2] while n is the principal quantum number. Inasmuch as n can take on only integer values $1, 2, 3, \ldots$, E is discrete rather than continuous. Although energy E depends only on the principal number n, the wave function ψ in general depends on a set of three quantum numbers: n, the principal quantum number, with permissible values $1, 2, 3, \ldots$; the orbital angular momentum quantum number l, with permissible values $0, 1, 2, \ldots, n-1$; and the magnetic quantum number, with permissible values $-l, -l+1, -l+2, \ldots, 0, \ldots, l-2, l-1, l$.

The Schrödinger equations, (4.1) and (4.2), do not include the existence of an electron spin. If this fact is included in the proper way, an additional quantum number m_s, the spin quantum number that can take on only values $\pm\frac{1}{2}$ will appear.[3] Based on these quantum numbers and the Pauli exclusion principle, which states that no two electrons can have the same set of quantum

---

[1]L. J. Schiff, *Quantum Mechanics*, McGraw-Hill, New York, 1955, p. 84.

[2]To obtain expression (4.3), it is necessary to replace m in Eq. (4.2) by $m_r = m_e M_p/(m_e + M_p)$ where m_e and M_p are, respectively, the electron and proton masses (see Schiff, *Quantum Mechanics*, p. 81). In this context, the proton can be considered as having infinite mass and being stationary while the electron orbits around it with effective mass m_r instead of m_e.

[3]P. A. M. Dirac, *The Principles of Quantum Mechanics*, Clarendon, Oxford, 1947, Ch. 2.

TABLE 4.1 ELECTRONIC STRUCTURE
OF SOME ELEMENTS

Element	Ground-state configuration
H	$1s$
He	$1s^2$
Li	$1s^2 2s$
Be	$1s^2 2s^2$
B	$1s^2 2s^2 2p$
C	$1s^2 2s^2 2p^2$
N	$1s^2 2s^2 2p^3$
O	$1s^2 2s^2 2p^4$
F	$1s^2 2s^2 2p^5$
Ne	$1s^2 2s^2 2p^6$
Na	$(1s^2 2s^2 2p^6)3s$
Mg	$(1s^2 2s^2 2p^6)3s^2$
Al	$(1s^2 2s^2 2p^6)3s^2 3p$
⋮	⋮
Ca	$(1s^2 2s^2 2p^6 3s^2 3p^6)4s^2$
Sc	$(1s^2 2s^2 2p^6 3s^2 3p^6)3d4s^2$
Ti	$(1s^2 2s^2 2p^6 3s^2 3p^6)3d^2 4s^2$
V	$(1s^2 2s^2 2p^6 3s^2 3p^6)3d^3 4s^2$
Cr	$(1s^2 2s^2 2p^6 3s^2 3p^6)3d^5 4s$
Mn	$(1s^2 2s^2 2p^6 3s^2 3p^6)3d^5 4s^2$
Fe	$(1s^2 2s^2 2p^6 3s^2 3p^6)3d^6 4s^2$
Co	$(1s^2 2s^2 2p^6 3s^2 3p^6)3d^7 4s^2$
Ni	$(1s^2 2s^2 2p^6 3s^2 3p^6)3d^8 4s^2$
Cu	$(1s^2 2s^2 2p^6 3s^2 3p^6)3d^{10} 4s$

numbers, the periodic table can be constructed. Consider the simplest possible atom, hydrogen; it consists of a proton and an electron. In the absence of excitation or thermal agitation, the electron would tend to occupy the lowest possible energy state, or ground state, corresponding to $n = 1$. According to the rules enumerated above, $l = 0$, $m_l = 0$, and $m_s = \pm\frac{1}{2}$ for this case. For the next heavier atom, helium, it follows that for one of the electrons $n = 1$, $l = 0$, $m_l = 0$, and $m_s = +\frac{1}{2}$, while for the other $n = 1$, $l = 0$, $m_l = 0$, and $m_s = -\frac{1}{2}$. In spectroscopic terms these two electrons are said to be in the $1s^2$ state.[4] Since the exclusion principle does not allow any other combination of l, m_l, and m_s for $n = 1$, the ground state of the next element, lithium, would be in the $1s^2 2s$ state, indicating that the third electron has the quantum number $n = 2$, $l = 0$, $m_l = 0$, and $m_s = \pm\frac{1}{2}$. The ground state of the elements in the periodic table up through copper are thus given as in Table 4.1.

It is interesting to note that before the shell $n = 3$ is filled, the $4s$ subshell is completely filled, as in calcium. The next element, scandium, begins to fill the empty $3d$ shell, and as we move up the periodic table through copper, we find that the $3d$ shell is gradually filled. These elements are referred to as transition

---

[4]In spectroscopic notation the 1 refers to $n = 1$ and s to $l = 0$. Accordingly, the ground state of hydrogen is designated $1s$, that of helium by $1s^2$, the superscript 2 indicating two electrons in the $1s$ state. Similarly, $l = 1, 2, 3, 4$ cases are designated by p, d, f, g, respectively.

elements. If we divide the $3d$ shell into two subshells, one representing positive electron spin and the other negative spin, we find that one subshell fills before any electron is added to the other subshell. Thus, the positive and negative spins are generally unbalanced, resulting in a large net magnetic moment associated with these elements. This peculiar behavior is governed by Hund's rule for the ground state of such atoms[5]:

1. The electron spins add in such a way as to give the maximum possible S consistent with the Pauli exclusion principle.
2. The orbital momenta combine in such a way as to give the maximum value of L that is consistent with rule 1.
3. For an incompletely filled shell, $J = L - S$ for a shell less than half occupied, and $J = L + S$ for a shell more than half occupied.

The problem of an atom with more than one electron is much more difficult to handle than the hydrogen atom because the electrons are in motion and their interaction energy depends on the instantaneous separation between them. Indeed, the many-electron problem can be treated analytically only with approximate methods. However, for ionized atoms each containing only one electron, such as He^+ and Li^{2+}, the energy and wave function are still hydrogenic in character. In this case the right-hand side of Eq. (4.3) should be multiplied by Z^2 where Z is the atomic number ($Z = 1$ for hydrogen, $Z = 2$ for helium, etc.).

4.2 DIAMAGNETISM

For atoms without a net spin, only the orbital motion of the electrons contributes to magnetism. Since the susceptibility of materials containing such atoms is usually negative, such materials are termed *diamagnetic*. We shall first give a simple classical derivation of the diamagnetic susceptibility.

An electron circulating in an orbit is analogous to a current flowing around a resistanceless loop. According to Lenz's law, if a magnetic field is applied to such a loop, a current will be induced in such a direction as to oppose the change in flux enclosed by the loop. In a similar manner, if a magnetic field H_0 is applied perpendicular to the plane of the orbit of an electron circulating about a proton with tangential velocity $\mathbf{v}$, the frequency of electronic circulation will change with H_0. From the Lorentz force law, $\mathbf{F} = q(\mathbf{E} + \mathbf{v} \times \mathbf{B})$ for a charge q, Maxwell's equation $\mathbf{V} \cdot \mathbf{D} = \rho$, and Gauss's theorem $\oint \mathbf{V} \cdot \mathbf{D} \, dV = \oint \mathbf{D} \cdot d\mathbf{S}$, we can easily show that the force acting on the electron due to the proton charge e and magnetic field H_0 is

$$\mathbf{F} = -\hat{\mathbf{a}}_r \frac{e^2}{4\pi\epsilon_0 r^2} + ev\mu_0 H_0 \tag{4.4}$$

[5]From studies on spectra, Hund arrived at these three rules that permit prediction of the magnetic moments of free atoms or ions in their ground state. S, L, and J are, respectively, the spin, orbital, and total angular momentum for an atom.

where $\hat{a}_r$ is a unit vector pointing from the proton toward the electron. This force is counterbalanced by the centrifugal force due to the electron's circulating motion. Thus, assuming the proton to be stationary for simplicity, we have

$$\frac{e^2}{4\pi\epsilon_0 r^2} + e\omega r\mu_0 H_0 = m\omega^2 r \tag{4.5}$$

where we have noted that $v = \omega r$. Solving for the frequency of circulation ω, we have from Eq. (4.5)

$$\omega = \omega_L \pm (\omega_L^2 + \omega_0^2)^{1/2} \tag{4.6}$$

where

$$\omega_L = \frac{e\mu_0 H_0}{2m} \tag{4.7}$$

$$\omega_0 = \left(\frac{e^2}{4\pi m\epsilon_0 r^3}\right)^{1/2} \tag{4.8}$$

It can be seen from Eqs. (4.5) and (4.8) that ω_0 is the frequency of rotation for zero H_0. Let us now consider two special cases represented by Eq. (4.6). First, if $\omega_L \gg \omega_0$, corresponding to the case of large H_0 or large r, to the first order in ω_0, ω is simply equal to $2\omega_L$ or ω_c, the well-known cyclotron frequency of a free electron in a magnetic field. On the other hand, if $\omega_L \ll \omega_0$, corresponding to the case of small H_0 or small r, to the first order in ω_L, ω is equal to $\omega_0 + \omega_L$. As anticipated earlier, the frequency of circulation is modified by the application of H_0, changing from ω_0 to $\omega_0 + \omega_L$ where ω_L is known as the Larmor frequency. Here the case represented by the minus sign in Eq. (4.6) has been discarded as nonphysical.

For the general case of $\mathbf{H}_0$ applied at an arbitrary angle to the plane of the orbit, the analysis is considerably more complicated. However, in diamagnetism we are mainly concerned with the motion of electrons strongly bound to the nucleus (r small) in the presence of a relatively weak magnetic field H_0. This case thus corresponds to the situation in which $\omega_L \ll \omega_0$. In other words, the electron will circulate about the proton many times before the plane of its orbit (no longer stationary as in the case where $\mathbf{H}_0$ is applied perpendicular to the orbit) precesses once about the magnetic field. Consequently, a magnetic moment $\boldsymbol{\mu}_c$ corresponding to electron circulation about a proton can be defined; $\boldsymbol{\mu}_c$ in turn can be considered as acted upon by the applied field $\mathbf{H}_0$. This very useful observation considerably simplifies our analysis.

Refer now to Fig. 4.1, where the normal to the electronic orbit makes a constant angle θ with the z axis and a variable angle ϕ (with respect to time t) with the x axis. According to classical mechanics, the torque $\boldsymbol{\tau}$ is equal to the rate of change of angular momentum $\mathbf{G}_c$ (where the subscript c designates circulation):

$$\boldsymbol{\tau} = \frac{d\mathbf{G}_c}{dt} \tag{4.9}$$

For the case in point, $\boldsymbol{\tau} = \mu_0 \boldsymbol{\mu}_c \times \mathbf{H}_0$ and $\mathbf{G}_c$ can be expressed in terms of $\boldsymbol{\mu}_c$, as will be seen. This and subsequent expressions assume that the relationship

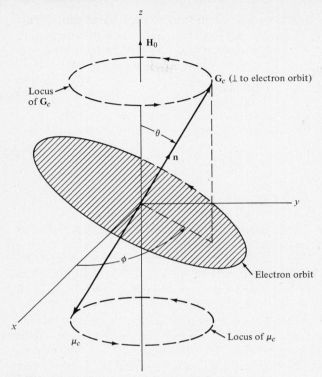

Figure 4.1 Precession about a magnetic field.

between **B**, **H**, and **M** is given by $\mathbf{B} = \mu_0(\mathbf{H} + \mathbf{M})$, consistent with tradition in magnetism. In this case **M** is considered **H**-like rather than **B**-like (for a discussion of units and conversion factors, see the Appendix). Since $\mathbf{G}_c = \hat{\mathbf{n}} I \omega$ where I is the moment of inertia of the orbit and $\hat{\mathbf{n}}$ is a unit vector perpendicular to the plane of the orbit with its direction defined by the right-hand rule (see Fig. 4.1). It follows that

$$\mathbf{G}_c = \hat{\mathbf{n}}(mr^2)\omega_0 \tag{4.10}$$

The magnetic moment $\boldsymbol{\mu}_c$, on the other hand, is equal to the product of current i due to electron circulation and area of the loop πr^2. Thus,

$$\boldsymbol{\mu}_c = -\hat{\mathbf{n}} \frac{e\omega_0}{2\pi} (\pi r^2) \tag{4.11}$$

and the minus sign appears because current flow corresponds to the flow of positive charges. Combining Eqs. (4.10) and (4.11), we easily find the sought-after relationship between $\mathbf{G}_c$ and $\boldsymbol{\mu}_c$:

$$\mathbf{G}_c = -\frac{2m}{e} \boldsymbol{\mu}_c \tag{4.12}$$

It immediately follows from Eq. (4.9) that the equation of motion for the orbital magnetic moment $\boldsymbol{\mu}_c$ is

$$\frac{d\boldsymbol{\mu}_c}{dt} = \gamma_c \boldsymbol{\mu}_c \times \mathbf{H}_0 \tag{4.13}$$

where

$$\gamma_c = -\frac{\mu_0 e}{2m} \tag{4.14}$$

is seen from Eq. (4.12) to be equal to $-\mu_0 |\boldsymbol{\mu}_c|/|\mathbf{G}_c|$. Substituting the expression

$$\boldsymbol{\mu}_c = (\hat{\mathbf{x}}\mu_{cx} + \hat{\mathbf{y}}\mu_{cy} + \hat{\mathbf{z}}\mu_{cz})e^{j\omega_L t} \tag{4.15}$$

into Eq. (4.13), we find that

$$\omega_L = \hat{\mathbf{z}}\gamma_c H_0 \tag{4.16}$$

From a consideration of Eqs. (4.13) and (4.14), we see that $\boldsymbol{\mu}_c$ precesses in the direction of increasing ϕ, as depicted in Fig. 4.1. In view of the definition of γ_c given by Eq. (4.14), we see that the expressions for ω_L given by Eqs. (4.7) and (4.16) are identical. This should clearly be so since the general expression [Eq. (4.16)] for the Larmor precession frequency should be applicable to the special case where $\boldsymbol{\mu}_c$ and $\mathbf{H}_0$ are parallel as well.

We now proceed to calculate the diamagnetic susceptibility. By definition, the susceptibility of a substance is defined as magnetic moment per unit volume (magnetization) per unit applied field. Thus, for this case only phenomena directly associated with the application of $\mathbf{H}_0$ are of real interest. In the context of the above discussion only the Larmor precession frequency ω_L, which is proportional to H_0, need be considered in susceptibility calculations. Accordingly, and in analogy with Eq. (4.11), the magnetic moment component in the direction of $\mathbf{H}_0$ due to the precessional motion of the electron, μ_{Lz}, is given by

$$\mu_{Lz} = -\frac{e\omega_L}{2\pi}(\pi\rho^2) \tag{4.17}$$

where ρ, as indicated in Fig. 4.1, is the projection of the radius vector $\mathbf{r}$ on the x-y plane. Since $\boldsymbol{\mu}_c$ rotates in the direction of increasing ϕ (see Fig. 4.1), the precessional angular momentum is in the $+z$ direction, while the precessional magnetic moment is in the $-z$ direction. This gives rise to a projection of $\boldsymbol{\mu}_L$ in a direction opposite to that of $\mathbf{H}_0$ and accounts for the negative sign in Eq. (4.17).

If there are Z electrons per atom and N atoms per unit volume (assume identical here), the diamagnetic susceptibility χ_d is given by

$$\chi_d = -\sum_{i=1}^{Z} \frac{N\mu_0 e^2}{4m}\pi\rho_i^2 \tag{4.18}$$

The summation $\sum_{i=1}^{Z} \rho_i^2$ can be set equal to $Z\overline{\rho^2}$ where $\overline{\rho^2} = \overline{x^2} + \overline{y^2}$ is the average of the square of the perpendicular distance of the electron from the field or z axis. If on average the distribution of electronic orbital orientations about the nucleus for a given atom is spherically symmetric, then $\overline{x^2} = \overline{y^2} = \overline{z^2}$.[6] Because $r^2 = x^2 + \overline{y^2} + \overline{z^2}$ is the mean square distance of the electron from the nucleus, it follows that $\overline{\rho^2} = \frac{2}{3}\overline{r^2}$. Accordingly, Eq. (4.18) becomes

$$\chi_d = -\frac{N\mu_0 Ze^2}{6m}\,\pi\overline{r^2} \tag{4.19}$$

χ_d is quite small, as can be estimated by substituting plausible values of N, Z, and $\overline{r^2}$ into Eq. (4.19). For example, if $N = 5 \times 10^{22}/\text{cm}^3$, $\overline{r^2} \simeq 10^{-16}\ \text{cm}^2$, and $Z = 10$, then χ_d is on the order of 10^{-5}. If one compares this value with that for paramagnets (10^{-4}–10^{-2}) or ferromagnets (10–10^4, say), one sees that it is very small indeed. We also note from Eq. (4.19) that χ_d is negative, indicating that the electronic orbits reorient themselves when an external field is applied in such a way as to oppose an increase in flux through its orbit. Thus, diamagnetism is present in all materials. However, in view of the relatively small magnitude of its susceptibility compared to that due to other types of magnetism, as mentioned above, it is significant only in materials without a net magnetic moment. Most free atoms and ions have an incompletely filled shell structure and therefore have a net magnetic moment. However, in practice, we are usually concerned with atoms in combination in which case the net magnetic moment is frequently zero. For example, in ionic compounds a positive ion loses as many electrons as necessary to a negative ion so that both ions have a resultant closed-shell structure and zero net moment. In covalent binding such as that existing in the hydrogen or nitrogen molecules H_2 and N_2, electrons are usually shared so that no moment exists in these cases either. In metals with zero net ion core spin, the diamagnetic susceptibility due to the ion core electrons is roughly given by Eq. (4.19). The diamagnetic contribution due to conduction electrons is more complicated. Furthermore, since conduction electrons also possess spin, they give rise to a paramagnetic susceptibility as well.

In the derivation of Eq. (4.19) we implicitly assumed that the direction of H_0 is an axis of symmetry of the system. In most molecular systems, however, this condition is not satisfied. For this general case the theory of Van Vleck shows that the susceptibility expression (4.19) is incomplete. Specifically, an additional positive term, identified as Van Vleck temperature-independent paramagnetism, appears.[7] If this term was large enough, the susceptibility can become positive rendering the material paramagnetic.

4.3 PARAMAGNETISM

Generally speaking, paramagnetism is important in substances in which the individual atoms have net magnetic moments but the coupling between the

---

[6]This is equivalent to the assumption of no net orbital magnetic moment.

[7]J. H. Van Vleck and A. Frank, *Proc. Natl. Acad. Sci. U.S.* **15**, 539 (1929).

magnetic moments of the various atoms is relatively weak. If an external magnetic field were applied to the substance, the individual moments would align with it if it were not for the randomizing effect of thermal agitation. At a given temperature T, we may therefore think of the magnetic moments as partially lining up with the external field, thus giving rise to a positive induced magnetic moment and a positive susceptibility.

First, let us examine the problem of paramagnetism in classical terms and then show that the result obtained for the paramagnetic susceptibility is the same as that derived from quantum mechanics. For an atom with a magnetic moment μ, the energy of interaction with the applied field $\mathbf{H}_0$ is

$$V = -\mu_0 \boldsymbol{\mu} \cdot \mathbf{H}_0 = -\mu_0 \mu H_0 \cos \theta \qquad (4.20)$$

where θ is the angle between $\boldsymbol{\mu}$ and $\mathbf{H}_0$. At thermal equilibrium, θ for the $\boldsymbol{\mu}$ of various atoms will take on a spectrum of values dependent upon the relative strengths of the Zeeman and thermal energies. It follows that the magnetization M (magnetic moment per unit volume) for a sample containing N atoms each with magnetic moment μ is simply given by

$$M = N\mu_0 \,\overline{\mu \cos \theta} \qquad (4.21)$$

where $\overline{\cos \theta}$ is the average value of $\cos \theta$ over a distribution in thermal equilibrium.

Since the interaction between the μs is assumed to be relatively weak in paramagnetism, their orientation distribution must obey Maxwell-Boltzmann statistics. According to statistical mechanics, the relative probability of finding a magnetic moment in an element of solid angle $d\Omega$ is proportional to $e^{-V/kT}$ where k is the Boltzmann constant.[8] Accordingly,

$$\overline{\cos \theta} = \int e^{-V/kT} \cos \theta \, d\Omega \Big/ \int e^{-V/kT} \, d\Omega \qquad (4.22)$$

Substituting Eq. (4.20) into Eq. (4.22) and integrating, we find

$$\overline{\cos \theta} = \coth a - \frac{1}{a} = L(a) \qquad (4.23)$$

where $a = \mu_0 \mu H / kT$ and $L(a)$ is known as the Langevin function.

Substituting Eq. (4.23) into Eq. (4.21), we obtain the expression for the magnetization M:

$$M = N\mu_0\mu L(a) = N\mu_0\mu \left(\coth \frac{\mu_0\mu H}{kT} - \frac{kT}{\mu_0\mu H} \right) \qquad (4.24)$$

where we have used the definition for a. It is seen from Eq. (4.24) that the paramagnetic susceptibility $\chi_p (= M/H)$ is in general a function of H. However, the available field strength in the laboratory is such that usually $\mu_0\mu H \ll kT$

---

[8]See, e.g., Aldert Van der Ziel, *Solid State Physical Electronics*, Prentice-Hall, Englewood Cliffs, NJ, 1957, p. 42.

even at low temperatures. In this case of very small a, $L(a)$ reduces to $a/3$ and the paramagnetic susceptibility is simply given by

$$\chi_p = \frac{N\mu_0^2\mu^2}{3kT} \tag{4.25}$$

Note that χ_p is positive as expected.

We now proceed to show that a quantum-mechanical derivation gives the same expression for χ_p, as does the classical result (4.25). To begin with, we have shown in Section 4.1 that each electron i of a given atom possesses an orbital angular momentum $\mathbf{l}_i\hbar$ and a spin angular momentum $\mathbf{s}_i\hbar$. These momenta in turn give rise to orbital and spin magnetic moments $\boldsymbol{\mu}_{li}$ and $\boldsymbol{\mu}_{si}$. To find the total magnetic moment of the atom it is necessary to sum up the magnetic moments, orbit and spin, of the various electrons. It turns out that there are two ways in which the various $\mathbf{l}_i\hbar$s and $\mathbf{s}_i\hbar$s can be summed quantum mechanically. The first is known as j-j coupling, in which the $\mathbf{l}_i\hbar$ and $\mathbf{s}_i\hbar$ of a given electron i add together to form a resultant total angular momentum for the electron of $\mathbf{j}_i\hbar$. The $\mathbf{j}_i\hbar$s of the various electrons in turn add together to form the resultant total angular momentum $\mathbf{J}\hbar$ for the atom. Mathematically, these statements can be summarized as follows:

$$\mathbf{J}\hbar = \sum_i \mathbf{j}_i\hbar = \sum_i (\mathbf{l}_i + \mathbf{s}_i)\hbar \tag{4.26}$$

For the second kind of coupling, known as the Russel-Saunders coupling, a corresponding relation can also be written:

$$\mathbf{J}\hbar = \mathbf{L}\hbar + \mathbf{S}\hbar = \sum_i \mathbf{l}_i\hbar + \sum_i \mathbf{s}_i\hbar \tag{4.27}$$

In other words, the $\mathbf{l}_i\hbar$s of the various electrons combine to form a total orbital angular momentum $\mathbf{L}\hbar$ and the $\mathbf{s}_i\hbar$s combine to form a total spin angular momentum $\mathbf{S}\hbar$. $\mathbf{L}\hbar$ and $\mathbf{S}\hbar$ in turn combine to form $\mathbf{J}\hbar$ for the whole atom. There is a temptation to say that the Js given by Eqs. (4.26) and (4.27) are the same based on the apparent mathematical equivalence of these equations. However, this is *not* true, because $\sum_i \mathbf{j}_i\hbar$, $\sum_i \mathbf{l}_i\hbar$, and $\sum_i \mathbf{s}_i\hbar$ may have to obey different quantum-mechanical selection rules.

It turns out that only in heavy elements such as uranium is the j-j coupling dominant. In materials of magnetic interest only the Russel-Saunders coupling is important. Accordingly, the corresponding atomic orbital and spin angular magnetic moments are given by

$$\mu_0\mu_L = \frac{\mu_0 e}{2m}\sqrt{L(L+1)}\,\hbar \tag{4.28}$$

$$\mu_0\mu_s = 2\frac{\mu_0 e}{2m}\sqrt{S(S+1)}\,\hbar \tag{4.29}$$

where we have used Eq. (4.12). According to the results of wave mechanics, the magnitude of $\mathbf{L}\hbar$ is $\sqrt{L(L+1)}\hbar$, $L\hbar$ being the projection of $\mathbf{L}\hbar$ along a given axis. Similarly, the magnitude of $\mathbf{S}\hbar$ is $\sqrt{S(S + 1)}\,\hbar$, $S\hbar$ being the projection of $\mathbf{S}\hbar$ along a given axis. Because of the negative charge of the electron, $\boldsymbol{\mu}_L$ and $\boldsymbol{\mu}_s$ are directed opposite to $\mathbf{L}\hbar$ and $\mathbf{S}\hbar$, respectively. Whereas $\mu_0|\boldsymbol{\mu}_L|/\sqrt{L(L + 1)}\hbar = \mu_0 e/2m$, experiments indicate that $\mu_0|\boldsymbol{\mu}_s|/\sqrt{S(S + 1)}\hbar = \mu_0 e/m$. Because of this difference of a factor of 2, the resultant magnetic moment $\mu_0\boldsymbol{\mu}_J$ will not be directed opposite to $\mathbf{J}\hbar$, as clearly demonstrated in Fig. 4.2. From consideration of the geometry, it can be shown that

$$\mu_0\mu_J = \frac{\mu_0 e\hbar}{2m}\, g\sqrt{J(J + 1)} \tag{4.30}$$

where the Landé g-factor is given by the expression

$$g = 1 + \frac{J(J + 1) + S(S + 1) - L(L + 1)}{2J(J + 1)} \tag{4.31}$$

If the atom is placed in a magnetic field $\mathbf{H}_0$ sufficiently weak that coupling between $\mathbf{L}\hbar$ and $\mathbf{S}\hbar$ is not broken down, then their resultant $\mathbf{J}\hbar$ will precess about the direction of $\mathbf{H}_0$ as an axis. The additional energy ΔE due to the interaction between $\boldsymbol{\mu}_J$ and $\mathbf{H}_0$ is simply

$$\Delta E = \mu_0\mu_J H_0 \cos\theta \tag{4.32}$$

where θ is the angle between $\boldsymbol{\mu}_J$ and $\mathbf{H}_0$. Substituting Eq. (4.30) into Eq. (4.32),

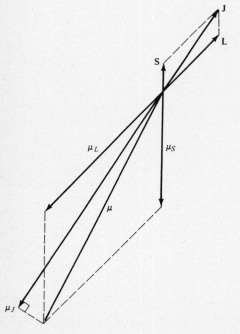

Figure 4.2 Magnetic moment and angular momentum vectors.

we have

$$\Delta E = \frac{\mu_0 e\hbar}{2m} gH_0 M_j \tag{4.33}$$

where $M_J(=\sqrt{J(J+1)} \cos \theta)$ is the projection of the vector $\mathbf{J}$ upon the field axis. For a given value of J, M_J can take on only $2J + 1$ discrete values.

Consider the simple example of free electrons in an applied field H_0. Since in this case $L = 0$ and $J = S$, $g = 2$ according to Eq. (4.31). Also $M_j = m_s = \pm\frac{1}{2}$ since the spin angular momentum of an electron along a given axis can only take on values of $\pm\frac{1}{2}\hbar$. It then follows from Eq. (4.33) that in this case

$$\Delta E = \frac{\mu_0 e\hbar}{2m} gH_0(1) \tag{4.34}$$

The quantity $\mu_0 e\hbar/2m$, designated by μ_B, is known as the Bohr magneton. If the electron has some reference energy E_0 in the absence of H_0, then this energy level will split into two nondegenerate levels, one located at $E_0 + \mu_B H_0$ and the other at $E_0 - \mu_B H_0$, as depicted in Fig. 4.3.

According to the Boltzmann distribution law, when there are only two energy levels in a magnetic field as for the case in point, the populations at these levels in thermal equilibrium are simply given by

$$\frac{N_1}{N} = \frac{e^{\mu_B H_0/kT}}{e^{\mu_B H_0/kT} + e^{-\mu_B H_0/kT}} \tag{4.35}$$

and

$$\frac{N_2}{N} = \frac{e^{-\mu_B H_0/kT}}{e^{\mu_B H_0/kT} + e^{-\mu_B H_0/kT}} \tag{4.36}$$

Here N_1 and N_2 are, respectively, the electron populations of the lower and upper energy levels, and $N = N_1 + N_2$ is the total number of electrons under consideration. Since the projection of μ along $\mathbf{H}_0$ for an electron is $-\mu_B$ in the upper state and $+\mu_B$ for the lower state, the resultant magnetization M for N atoms per unit volume is

$$M = (N_1 - N_2)\mu_B = N\mu_B \tanh\left(\frac{\mu_B H_0}{kT}\right) \tag{4.37}$$

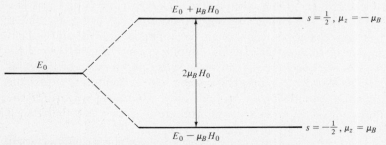

Figure 4.3 Splitting of energy levels in a magnetic field.

where Eqs. (4.35) and (4.36) have been used. For ordinary conditions in the laboratory, $\mu_B H_0 \ll kT$ and $\tanh(\mu_B H_0/kT)$ reduces to $\mu_B H_0/kT$. Accordingly, the quantum-mechanical paramagnetic susceptibility is simply given by

$$\chi_p = \frac{N\mu_B^2}{kT} \tag{4.38}$$

To compare this expression with that given by Eq. (4.25) and derived classically, it is necessary to establish the relationship between μ (corresponding to the magnitude of $\boldsymbol{\mu}$) and μ_B (corresponding to the projection of $\boldsymbol{\mu}$ along $\mathbf{H}_0$). According to wave mechanics, the magnitude of $\mathbf{S}$ is $\sqrt{S(S+1)} = (\sqrt{3}/2)S$ as $S = \frac{1}{2}$. Since the ratio between the magnitudes of $\boldsymbol{\mu}$ and $\mathbf{S}$ is a constant, we conclude accordingly that $\mu_0\mu = \sqrt{3}\mu_B$. Thus, we see that the classical result for χ_p given by Eq. (4.25) is identical to the quantum-mechanical result given by Eq. (4.38).

A similar but somewhat more complicated derivation can be carried out for the case of an atom. If the atom has a total angular momentum $J\hbar$, it will have $2J + 1$ equally spaced energy levels in a magnetic field corresponding to $M_J = -J, -J + 1, \ldots, 0, \ldots, J + 1, J$. In this general case it can be shown that

$$M = NgJ\mu_B B_J(x) \tag{4.39}$$

with the Brillouin function $B_J(x)$ given by the expression[9]

$$B_J(x) = \frac{2J+1}{2J}\coth\left(\frac{2J+1}{2J}\right)x - \frac{1}{2J}\coth\frac{x}{2J} \tag{4.40}$$

where $x = gJ\mu_B H_0/kT$. Again, for ordinary laboratory conditions, $x \ll 1$ and the paramagnetic susceptibility for N atoms per unit volume is simply given by

$$\chi_p = \frac{C}{T} \tag{4.41}$$

where

$$C = \frac{NJ(J+1)g^2\mu_B^2}{3k} \tag{4.42}$$

is known as the Curie constant and Eq. (4.41) is known as the Curie law. As mentioned previously, χ_p ranges from 10^{-4} to 10^{-2}, a number which although large compared to the diamagnetic susceptibility $x_d(\sim -10^{-5})$ is still very small compared to the susceptibility of ferromagnetic materials $(10-10^4)$.

A number of rare-earth ions, such as Gd^{3+}, actually obey the Curie law very well. For salts of the iron transition group, however, good agreement between theory and experiment can often be obtained only if $gJ(J + 1)$ is replaced by $2S(S + 1)$. In view of the definition of g given by Eq. (4.31) and the relation $\mathbf{J} = \mathbf{L} + \mathbf{S}$, we conclude that in these salts, unlike those of the rare-earth group, the orbital contribution is for some reason unimportant and the magnetic moment is essentially due to spin only. The basic reason for the difference in behavior of the rare-earth and iron group salts is that the $4f$ shell

---

[9]Note that as J becomes very large, corresponding to the case of a macroscopic sample, $B_J(x) \to L(a)$ given by Eq. (4.23) for $g = 2$, $J = S = \frac{1}{2}$, i.e., for a free electron, as expected.

(where there is a net magnetic moment), responsible for paramagnetism in the rare-earth ions, lies deep inside the ions, being partly shielded from the environment by the outer $5s$ and $5p$ shells, whereas in the iron group, the $3d$ shell, responsible for paramagnetism, is the outermost shell in the ionic state. The $3d$ shell is thus exposed to the intense local electric fields due to neighboring ions and the dipole moment of water by hydration in the crystal. The interaction of the paramagnetic ions with the crystalline electric field may have an important effect on the lack of contribution of the orbital motion of the electrons to the magnetic moment, as discussed below.

If an electric field is directed toward (or away from) a fixed center such as a nucleus, the plane of the classical orbit about the center will remain fixed in space, so that the orbital angular momentum components L_x, L_y, and L_z are constant with time. If an inhomogeneous electric field such as the crystalline field is superimposed upon the central field, however, the orientation of the plane of the orbital will change with time. Under these circumstances, the angular momentum components are no longer constant and may average to zero. In this case the motion of the electron is said to be quenched, leading to the conclusion that $J = S$ and $g = 2$.[10] Apparently, quenching does occur in salts of the iron transition group and as a matter of fact in transition metals (discussed in Section 4.4) as well. When the spin-orbit interaction energy is introduced as an additional perturbation in the system, the quenching may be partially lifted because the spin may carry some orbital momentum with it, and the g value as given by Eq. (4.31) will differ slightly from 2.

Because of its relatively small susceptibility, paramagnetic materials are not, in contrast to the case of ferromagnetic materials, extensively used. However, it is used in the maser, an extremely low noise amplifier that is utilized, among other things, in radio astronomy to detect extraordinary weak signals from the heavens.[11]

4.4 FERROMAGNETISM

In our study of paramagnetism we implicitly assumed that the interaction between adjacent atomic magnetic moments is sufficiently weak so that in the absence of an external field, these moments are randomly oriented due to the everpresent thermal agitation. In ferromagnetic materials, however, this is not necessarily true. It turns out that in this class of materials, the interaction between adjacent atomic magnetic moments is very large indeed. In fact, the interaction is so very large that in most of these materials atomic moments are very nearly aligned even at room temperature and beyond. The origin of this strong interaction is quantum-mechanical rather than classical as we shall justify below.

---

[10]In quantum mechanics L^2 and L_z instead of L_x, L_y, and L_z are constants of motion in a central field. To a good approximation, L^2 is still a constant in the presence of this inhomogeneous electric field, but L_z need not be; if L_z averages to zero, the motion is said to be quenched.

[11]R. F. Soohoo, *Microwave Electronics*, Addison-Wesley, Reading MA, 1971, p. 237.

If the atomic moments μ were to be aligned, the energy of interaction between them must be at least equal to the energy of thermal agitation at temperature T. If we characterize the interaction by an effective field H_e, then the following near-equality must hold:

$$\mu H_e \simeq kT \tag{4.43}$$

If μ is set equal to μ_B corresponding to one net spin per atom, H_e is found to be equal to 6×10^4, 4.5×10^6, and 2.1×10^7 Oe for $T = 4°$, $300°$ and $1388°K$, respectively. These temperatures correspond successively to liquid helium temperature, room temperature, and the Curie temperature of cobalt, the highest T_c for a ferromagnetic material. As we shall show later, the Curie temperature T_c of a ferromagnetic material is the temperature at which atomic magnetic moment alignment is destroyed by thermal agitation.

We see from the above calculation that even at liquid helium temperatures some 60,000 Oe are required to align the atomic moments. For ferromagnetic materials with T_c at room temperature or higher, more than a million oersteds are required to keep the moments parallel. If we compare this to 200,000 Oe, the highest value of static magnetic field ever produced in any laboratory, we easily realize the enormous strength involved. The origin of this tremendous field is therefore unlikely to be classical in nature, due to classical dipole-dipole forces, say. That this is so can best be appreciated by examining the magnitude of the field acting on a dipole in the direction of the magnetization M in a magnetized medium; known as the Lorentz field, it is equal to $M/3$, as we shall see in Section 4.6. Even for iron, which has a saturation magnetization ($4\pi M$ at $T = 0$) of 21,850 G and a T_c of $1043°K$, this field is only 7280 Oe, orders of magnitude too small to account for the occurrence of ferromagnetism at ordinary temperatures. We conclude, therefore, that if classical effects were dominant, the interaction between atomic moments in ferromagnetic materials would be insignificant, even at extraordinarily low temperatures.

The quantum-mechanical origin of the effective field, called the *exchange field*, in ferromagnetic materials lies in the interplay between the Coulomb energy and the Pauli exclusion principle. Although the effect can be represented by an effective magnetic field, we must emphasize that it is *electrostatic* rather than magnetostatic in origin. In a solid, magnetostatic forces due to magnetic dipoles are much smaller in magnitude than electrostatic forces due to electric charges.

Although the Lorentz field is far from sufficient to account for the enormous effective field in ferromagnetic materials, it is directed in the proper direction, that is, in the direction of M. In other words, the direction of the Lorentz field is such as to cause the parallel alignment of atomic moments. One is tempted, therefore, to assume that the effective field can be represented by the term λM where λ ($\gg \frac{1}{3}$) is characteristic of a given ferromagnetic material. Accordingly, the total field H can be expressed in terms of the applied field H_0 and the effective field λM as[12]

$$H = H_0 + \lambda M \tag{4.44}$$

[12]P. Weiss, *J. Phys.* **6**, 667 (1907).

In this expression we have implicitly assumed that the medium is saturated in the direction $\mathbf{H}_0$, i.e., with $\mathbf{M} \parallel \mathbf{H}_0$, so that $\mathbf{H}$, $\mathbf{H}_0$ as well as $\lambda \mathbf{M}$ are in the direction of $\mathbf{M}$. Note also that we have neglected field components of classical origin, i.e., demagnetizing and Lorentz fields, in comparison with the Weiss molecular field λM; H_0 is retained, however, in order to allow us to calculate the ferromagnetic susceptibility $\chi_f = M/H_0$.

If the temperature T of the ferromagnetic material is above the Curie temperature T_c, the alignment between atomic moments will be destroyed. Under these circumstances, the material should behave much like a paramagnet. Accordingly, M/H should be equal to C/T; combining Eqs. (4.41) and (4.44) we thus have

$$\frac{M}{H_0 + \lambda M} = \frac{C}{T} \tag{4.45}$$

Solving for M/H_0, we find

$$\chi_f = \frac{C}{T - T_c} \tag{4.46}$$

where $T_c = C\lambda$ is the Curie temperature. We see from this expression and Eq. (4.41) that χ_f has the same form as χ_p when plotted against T except that the singularity occurs at $T = T_c$ rather than at $T = 0$, as depicted in Fig. 4.4. Relation (4.46) is known as the Curie-Weiss law.

We must turn our attention next to the temperature region below T_c. As $T \to 0$, the Weiss molecular field becomes increasingly effective in holding the atomic moments parallel. Inasmuch as $\lambda M \gg H_0$, H_0 will have little influence on the alignment of $\mathbf{M}$ below T_c. This can easily be verified by first rewriting

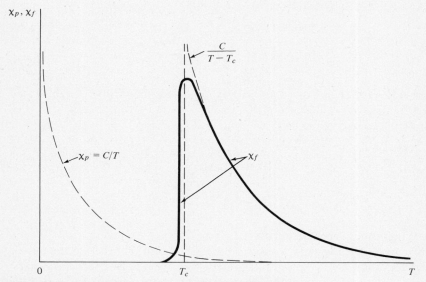

Figure 4.4 Susceptibility as a function of temperature.

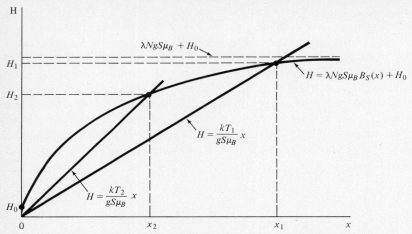

Figure 4.5 H vs x diagram for the determination of M vs T characteristics.

Eq. (4.39) for the magnetization of a material containing N atoms per unit volume:

$$M = NgS\mu_B B_s(x) \tag{4.47}$$

with
$$x = \frac{gSu_B H}{kT} \tag{4.48}$$

In these expressions we have replaced J by S since ferromagnetism, as discussed previously, is mainly due to the existence of electron spins. The equation for $B_s(x)$ is the same as that for $B_J(x)$ given by Eq. (4.40) except that J should be replaced by S. To proceed further, we must rearrange both Eqs. (4.47) and (4.48) so that H is exhibited in terms of x. With the help of Eq. (4.44), we have

$$H = \lambda NgS\mu_B B_s(x) + H_0 \tag{4.49}$$

and
$$H = \frac{kT}{gS\mu_B} x \tag{4.50}$$

We can now plot H in Eqs. (4.49) and (4.50) as a function of x, as illustrated in Fig. 4.5, for given values of λ, S, H_0, and T—say, T_1. We note that at $x = x_1$ the straight line and curve cross each other, thus determining the permissible value of H at T_1, i.e., H_1. In other words, at a given T_1 there is some value of H_1 that makes the right-hand sides of Eqs. (4.49) and (4.50) equal. This value of H_1 in turn gives an M_1, which, according to Eq. (4.44), is

$$M_1 = \frac{H_1 - H_0}{\lambda} \tag{4.51}$$

If this procedure is repeated for other values of T (see, e.g., point x_2, H_2 corresponding to T_2 on Fig. 4.5), we can eventually obtain the M-T characteristic for various values of H_0. The result is shown in Fig. 4.6. In order to obtain a universal curve for the spontaneous magnetization (i.e., for the $H_0 = 0$ state)

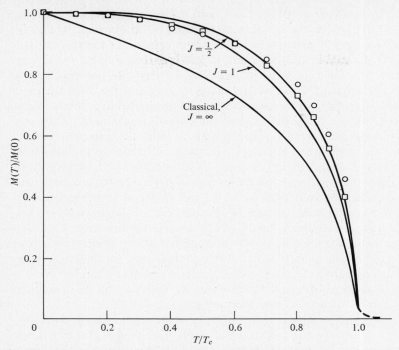

Figure 4.6 Normalized magnetization vs normalized temperature. (Circles: Fe; squares: Co, Ni.) [See F. Tyler, *Philos. Mag.* 11, 596 (1931).]

characteristic of a given value of S, we have plotted $M(T)/M(0)$ vs T/T_c where $M(T)$ is the magnetization at any temperature T and $M(0)$ the magnetization at $T = 0$. $M(0)$, corresponding to the maximum possible value of M, is referred to as the *saturation magnetization*. $M(0)$ is equal to $NgS\mu_B$ according to Eq. (4.47) since $B_s(\infty) \rightarrow 1$; this value corresponds to the case in which all atomic moments are parallel. Two cases are shown in Fig. 4.6, corresponding to $H_0 = 0$ and $H_0 \neq 0$; S is set equal to $\frac{1}{2}$, 1, or ∞. As expected, the $M(T)/M(0)$ vs T/T_c curve remains essentially unchanged when a finite H_0 is applied except in the region very close to T_c; for $T > T_c$, a slight tail appears. Experimental points for Ni, Fe, and Co are plotted on this same figure, and we can see that they exhibit the general behavior predicted by theory. The degree of agreement between theory and experiment depends, however, on the value of S assigned to these elements. According to Table 4.1, Ni, Fe, and Co in their free ionic state have respectively 2, 4, and 3 net spins, giving rise to 2, 4, and 3 Bohr magnetons of magnetic moment per atom or, equivalently, an S of 1, 2, and 1.5, respectively. Actually, the situation is not quite so simple once these ions are brought together to form a solid; in this case interaction between ions may change the effective magnetic moment per atom. Experimentally observed magnetic moment per atom is usually considerably smaller than the theoretical values calculated for the free ions and are also frequently not integral multiples of a Bohr magneton. For Ni, Fe, and Co the effective atomic magnetic moment has been experimentally found to be 0.606, 2.221, and 1.716, respectively.[13] Although these results can be explained by allowing mixtures of various ionicity, perhaps the most natural way of accounting for the nonintegral magnetic moment is to abandon the ionic model altogether and to adopt instead a band or collective electron model. In this model the $3d$ electrons, for example, are visualized as being in two energy bands, one with electron spin up and the other with spin down. If follows that the net spin would depend on the relative electron occupation of these bands. Since the $3d$ shell can contain a maximum of ten electrons, the spin-up and spin-down bands each contain five states per atom. If there are n_+ electrons in the spin-up band and n_- electrons in the spin-down band, then the net magnetic moment per atom is $n_+ - n_-$ Bohr magnetons. Thus, both the smaller-than-expected and noninteger atomic magnetic moments observed can be accounted for by a band theory of ferromagnetism.

In ferromagnetic materials like Ni, Fe, and Co the $3d$ electrons are very likely to be neither completely localized nor completely itinerant. Thus, it is unlikely that either the ionic model or the band model can completely explain all the observed results in these materials. For this reason, we shall proceed no further in the comparison of the theoretical and experimental results shown in Fig. 4.6 except to emphasize again that despite the ambiguity in the assignment of S, they agree with each other in their general features.

We now return to a more detailed study of the effect of an applied magnetic field H_0 on a ferromagnetic material. If $\mathbf{H}_0$ is applied in the direction of $\mathbf{M}$, we see from Fig. 4.6 that M changes very little due to the application of H_0 except at temperatures near or above the Curie temperature. Accordingly, if we define a ferromagnetic susceptibility χ_f, it should be given by

$$\chi_f = \frac{(M)_{H_0 \neq 0} - (M)_{H_0 = 0}}{H_0} \tag{4.52}$$

From the general feature of the $H_0 \neq 0$ curves in Fig. 4.6, we can sketch χ_f for $T < T_c$ as shown in Fig. 4.4; in drawing this portion of the curve, we have allowed for a smooth transition to x_f in the paramagnetic region given by the Curie law.

The situation becomes very different indeed when $\mathbf{H}_0$ is applied at an angle to $\mathbf{M}$ of a ferromagnet at $T < T_c$. Here the component of M in the direction of H_0, M_z increases with but is not necessarily proportional to H_0. Furthermore, M_z is a multiple-valued function of H_0; the value of M_z attained at a given H_0 depends on the history of H_0 to which the material has been subjected. Strictly speaking, $M_z(t)$ depends on $H_0(t)$ for t extending from $-\infty$ to t. Although under these circumstances the concept of a ferromagnetic susceptibility is still useful, it is necessary to consider also the B-H_0 or M-H_0 loops in their totality. We defer the study of these loops and associated ferromagnetic phenomena until we have considered the existence of domains in ferromagnetic materials (Section 4.6).

---

[13]These values were deduced from magnetization measurements with M equated to the product of atoms per volume and the effective magnetic moment per atom.

4.5 EXCHANGE INTERACTION

In Section 4.4 we examined the macroscopic theory of ferromagnetism in some detail. We concluded that a million or more oersteds of effective field must exist inside the material in order to account for the high Curie temperature in ferromagnetic materials. We further showed that such enormous fields cannot be of classical origin. Consequently, we must now turn to a study of the quantum-mechanical origin of ferromagnetism.

In 1928 Heisenberg showed that the Weiss molecular field λM is caused by quantum-mechanical exchange forces. As a consequence of the Pauli exclusion principle in quantum mechanics, it is usually not possible to change the relative direction of two spins without at the same time changing the spatial charge distribution in the overlapping region. The resulting change in the Coulomb energy of interaction between the spin-bearing charges, except for a spin-orientation-independent term, can be shown to be given by the expression[14]

$$E_{ex} = -2J_{ij}\mathbf{S}_i \cdot \mathbf{S}_j \tag{4.53}$$

where E_{ex} is known as the exchange energy and J_{ij} is the exchange integral between spins $\mathbf{S}_i$ and $\mathbf{S}_j$ of the charge distributions i and j.[15] Recalling the relation $s\hbar = -(m/\mu_0 e)\boldsymbol{\mu}_s$, according to the discussion following Eqs. (4.28) and (4.29), we can change Eq. (4.53) to read

$$(E_{ex})_i = -\boldsymbol{\mu}_i \cdot (\mathbf{H}_{ex})_j \tag{4.54}$$

where
$$(\mathbf{H}_{ex})_j = 2J_{ij}\left(\frac{m}{\hbar\mu_0 e}\right)^2 \boldsymbol{\mu}_j \tag{4.55}$$

Here $(\mathbf{H}_{ex})_j$ can be thought of as the field acting on magnetic moment $\boldsymbol{\mu}_i$ due to magnetic moment $\boldsymbol{\mu}_j$; in this context $(E_{ex})_i$ is equivalent to the Zeeman energy of interaction between a magnetic moment $\boldsymbol{\mu}_i$ and applied field $(\mathbf{H}_{ex})_j$. In other words, the form of Eq. (4.53) makes it appear as if there were a direct magnetic coupling between spins $\mathbf{S}_i$ and $\mathbf{S}_j$ although E_{ex} is electrostatic rather than magnetostatic in origin. It should again be emphasized, however, that E_{ex} has *no* classical analog.

If there are N spins per unit volume, the exchange energy density e_{ex} is given by

$$e_{ex} = -2 \sum_{i,j=1}^{N}{}' J_{ij}\mathbf{S}_i \cdot \mathbf{S}_j \tag{4.56}$$

where the prime indicates that the case $i = j$ accounting for the spin self-energy should be excluded. In summing the terms in Eq. (4.56), written in its present

[14]See, e.g., D. Bohm, *Quantum Theory*, Prentice-Hall, Englewood Cliffs, NJ, 1951, p. 490.

[15]Quantum mechanically, Eq. (4.53) should really be an operator equation, written $\mathcal{H}_{ex} = -2J_{ij}S_iS_j$ where $\mathcal{H}_{ex}$, known as the exchange Hamiltonian, and S_i, S_j are all operators. However, for many purposes in ferromagnetism it is a good approximation to treat S_i and S_j as classical vectors $\mathbf{S}_i$ and $\mathbf{S}_j$.

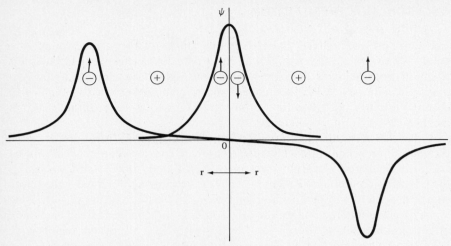

Figure 4.7   Electron wave functions in a hydrogen molecule.

form, care must be taken to sum over any pair of spins only once.[16] In general, J_{ij} decreases very rapidly with the separation between charge distributions i and j since its value is determined by the degree of their overlap. In practice, therefore, only nearest-neighbor terms are considered in the summation of Eq. (4.56). For example, if there are Z equally spaced neighbors for each and every spin in the sample and the spins are all parallel, $e_{ex} = -2N(ZJS^2)$ and an $\mathbf{H}_{ex}$ acting on each spin equal to $2ZJ(m/\hbar\mu_0 e)^2\mu$, according to Eq. (4.55).

The interplay between Coulomb interaction and spin direction correlation can be appreciated by looking at the simple hydrogen molecule. There are two eigenstates for the electrons with spatial wave functions ψ shown in Fig. 4.7. It is clear from this picture that since electrons in a singlet state have a high probability of being very close together, their spins must be oppositely-directed if Pauli's exclusion principle is not to be violated; if their spin orientations were the same, they would possess the same set of quantum numbers. We note that the total wave function of the singlet state, which in the absence of spin-orbit interaction is simply the product of spatial and spin wave functions, is anti-symmetric. In fact, the antisymmetrization of the total wave functions is a general requirement in quantum mechanics as a consequence of the Pauli exclusion principle. For the same reason, the electron spins in the triplet state must be oriented in the same direction since the spatial wave function in this case is already antisymmetric.

In the singlet state the two electrons are likely to be very close together and situated between the protons, while in the triplet state they are likely to be relatively far apart. Consequently, the Coulomb energy between the various particles in a hydrogen molecule must be different for the two different states

[16]Alternatively, this sum can be carried out exactly as shown in Eq. (4.56), thus summing the energy of each pair twice. In that case the factor of 2 in Eq. (4.56) should be dropped.

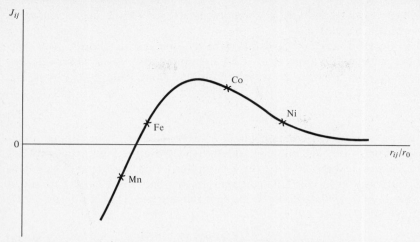

Figure 4.8 Exchange integral J_{ij} vs interionic distance r_{ij} divided by the orbital radius of $3d$ electrons.

since it is dependent upon the separation between particles. It is precisely this difference between the Coulomb energies of the triplet and singlet states of the molecule that gives rise to an exchange energy of the form given by Eq. (4.53). In order to attain minimum energy, the $\mathbf{S}_i$ and $\mathbf{S}_j$ of Eq. (4.56) must be parallel if the exchange integral J_{ij} is positive. This corresponds to the case of ferromagnetism in which the net atomic spins are aligned parallel. On the other hand, if J_{ij} is negative, antiparallel spin orientation, or antiferromagnetism, results. In the ferromagnetic case there is, of course, a resultant spontaneous magnetization, whereas in the antiferromagnetic case there is no net magnetic moment for the sample as a whole since there are equal number of spins oriented in opposite directions.

In general, J_{ij} is negative, favoring the nonferromagnetic state. However, a qualitative analysis by Bethe shows that J_{ij} is likely to be positive if the distance r_{ij} between the nuclei of the interacting ions i and j is fairly large compared to the orbital radius r_0 of the relevant electrons involved, as depicted in Fig. 4.8.[17] According to Slater, the ratio r_{ij}/r_0 should be larger than 3, but not much larger.[18] It may be seen from Fig. 4.8 that the transition elements Ni, Fe, and Co have positive exchange integrals, leading to ferromagnetic behavior, as is well known. On the other hand, Mn has a negative J_{ij} and is therefore antiferromagnetic, also in accord with experiment.

It would be instructive to establish a relationship between the exchange integral J_{ij} and the Curie temperature T_c. According to Eqs. (4.54) and (4.55), the larger the exchange integral, the larger the exchange field and exchange energy. A higher exchange energy in turn implies an ability to hold the spins parallel (or antiparallel) up to a higher temperature. We therefore expect the Curie

[17]W. Shockley, *Bell Syst. Tech. J.* **18**, 719 (1938).

[18]J. C. Slater, *Phys. Rev.* **36**, 57 (1930).

temperature at which the spin alignment is destroyed by thermal agitation to increase with increasing J_{ij}.

Let the atom under consideration have Z nearest neighbors, each connected with the central atom by an exchange integral J. If we neglect the exchange interaction between the central atom and more distant neighbors, Eq. (4.56) becomes

$$E_{ex} = -2JZS^2 \tag{4.57}$$

where we have neglected components of **S** perpendicular to the average magnetization. This is fully justifiable provided the temperature T is sufficiently less than T_c. However, when $T \to T_c$, the assumption of no transverse component of **S** is clearly not accurate. Thus, the relationship between T_c and J obtained below should be considered only as an approximate one. Combining Eq. (4.57) with Eq. (4.34), we have

$$2JZS^2 = gS\mu_B\lambda M \tag{4.58}$$

where we have replaced H_0 in Eq. (4.34) by the Weiss molecular field λM. The magnetization M is equal to the number of atoms per unit volume N times $gS\mu_B$, again provided T is sufficiently less than T_c as can be verified by means of Eq. (4.47) and related definitions (4.40) and (4.48). Substituting this expression for M into Eq. (4.58), we finally arrive at the relationship between $T_c = C\lambda$ where C is the Curie constant given by Eq. (4.42) and J:

$$T_c = \frac{2ZS(S + 1)}{3k} J \tag{4.59}$$

Indeed, as anticipated, the Curie temperature increases with the strength of the exchange coupling between spins.

4.6 FERROMAGNETIC DOMAINS

In a ferromagnetic material such as a piece of iron that has never been subjected to an applied magnetic field, the magnetization along any given direction is zero. On the other hand, we have just shown that the magnetic moments of a ferromagnetic material are aligned parallel, so a net magnetization in a sample that has never been subjected to an applied field should be expected. This seeming contradiction can be resolved if we postulate that in the demagnetized state the ferromagnetic material is divided up into many randomly oriented domains within each of which the magnetic moments are all aligned in one direction.[19] Such division is justified on the grounds that a ferromagnetic material composed of many domains is, under certain circumstances, in a state of lower energy than one with a single domain. Indeed, the size and shapes of these domains are determined by the condition of minimum energy for the sample in question.

[19]P. Weiss, *J. Phys.* **6**, 661 (1907).

This energy may be decomposed into a number of components arising in different ways inside the material. It has been shown that in the absence of an external applied field a condition of equilibrium, for which the sum of these energy components are at a minimum, is realized when domains have certain sizes and geometrical shapes. The predicted shapes have in fact since been found to occur; with suitable technique they can be mapped and viewed under optical and electron microscopes. To understand these patterns more fully, it is now necessary to consider the various energy components alluded to.

The energy in a ferromagnet can be conveniently divided into four components, each with an identifiable origin: (1) anisotropy energy, (2) magnetostatic energy, (3) magnetostrictive energy, and (4) domain wall energy. We shall discuss each of these energies briefly, followed by a simple calculation to illustrate their role in ferromagnetic domain formation.

4.6.1 Anisotropy Energy

Most materials of magnetic interest have a cubic crystal structure. Experimentally, it has been found that the free-energy density F of a ferromagnetic cubic single crystal sample can be represented quite well by the expression

$$F = K_0 + K_1(\alpha_1^2\alpha_2^2 + \alpha_2^2\alpha_3^2 + \alpha_3^2\alpha_1^2) + K_2\alpha_1^2\alpha_2^2\alpha_3^2 \tag{4.60}$$

where K_0, K_1, and K_2 are, respectively, the zeroth-, first-, and second-order anisotropy constants and α_1, α_2, and α_3 are direction cosines referred to the cube edges. Indeed, Eq. (4.60) can easily be shown to be consistent with the requirement of symmetry imposed by the periodicity of the crystal.[20] Since K_0 is independent of the direction cosines, the anisotropy energy density e_k is given by the K_1 and K_2 terms. K_1 and K_2 are in general functions of temperature. At room temperature, $K_1 = 4.2 \times 10^5$ erg/cm^3 and $K_2 = 1.5 \times 10^5$ erg/cm^3 for iron, whereas $K_1 = -5 \times 10^4$ erg/cm^3 for nickel. In general, K_1 is substantially larger than K_2; thus, the difference in sign between the K_1 of iron and nickel is significant. Indeed, an examination of Eq. (4.60) shows that e_k (or F) is minimum along the cube edges for iron and along the body diagonal for nickel. Correspondingly, e_k is maximum along the body diagonal for iron and along the cube edges for nickel. Accordingly, the cube edges of iron and body diagonal of nickel are designated as easy directions. Conversely, the body diagonal of iron and the cube edges of nickel are known as hard directions. If anisotropy energy were the only component of consequence in determining the magnetization direction, then **M** would lie along the cube edges of an iron single crystal and along the body diagonal of a nickel single crystal when no external field is applied. In order to move **M** away from these directions, it is necessary to supply some additional amount of energy. The energy required to move the magnetization from the easy to the hard direction in each case is known as the crystalline anisotropy energy of the crystal.

[20]For a more detailed discussion of anisotropy, see, e.g., R. F. Soohoo, *Magnetic Thin Films*, Harper & Row, New York, 1965, Ch. 7.

Cobalt, unlike iron and nickel, has a hexagonal crystal structure. For these types of crystals, e_k is given by

$$e_k = K'_1 \sin^2 \theta + K'_2 \sin^4 \theta \tag{4.61}$$

where θ is the angle that $\mathbf{M}$ makes with the hexagonal axis. At room temperature $K'_1 = 4.1 \times 10^6$ erg/cm^3 and $K'_2 = 1.0 \times 10^6$ erg/cm^3 for cobalt.

In many cases the origin of anisotropy can be attributed to spin-orbit coupling. As discussed in Sections 4.3 and 4.4, the orbital angular momenta of electrons are strongly but not completely quenched by the electrostatic field of other ions as they are brought together to form a solid. Since the orbital motion of the electrons is influenced by the electrostatic fields set up by the crystal lattice and the overlapping of the wave functions of neigboring atoms, the electron spin can "sense" the structure of the crystal via the agency of spin-orbit coupling. Owing to the discrete nature of the lattice, the electrostatic field seen by the electron spin may be a function of its orientation with respect to the crystalline axes. For this reason, it is plausible to say that there may be certain preferred directions of electron spin orientation for which the anisotropy energy is minimum. These directions correspond, of course, to the easy directions of magnetization of the crystal.

It is instructive to point out here that neither the exchange interaction between electron spins nor the ordinary dipole-dipole interaction due to spin magnetic moments adequately account for the observed anisotropy in ferromagnetic crystals. In the former case we see from Eq. (4.53) that the energy between two spins is dependent on the angle between them but is completely independent of the angle between the spins and the crystalline axes. In the latter case the ordinary dipole-dipole interaction can be represented by[21]

$$H_z = \sum_i \frac{3\mu_i z_1^2 - \mu_i r_1^2}{4\pi r_i^5} \tag{4.62}$$

where H_z is the z component of magnetic field due to an assembly of magnetic moments μ_i along the z axis at points designated by the radius vector $\mathbf{r}$, and $\mathbf{z}$ is the projection of $\mathbf{r}$ along the z axis. If the crystal is cubic, we have from symmetry,

$$\sum_i \frac{z_i^2}{r_i^5} = \sum_i \frac{x_i^2}{r_i^5} = \sum_i \frac{y_i^2}{r_i^5}$$

so that $\sum_i (r_i^2/r_i^5) = \sum_i 3(z_i^2/r_i^5)$ and H_z is identically zero. Thus, if the atoms can be represented by *parallel* magnetic dipole moments as is justified for a sample at low temperatures where thermal agitation is small, then the magnetic field acting on a given magnetic moment due to all the others is zero. Consequently, if magnetic anisotropy were due to dipole-dipole interaction, it would approach zero as $T \to 0$. Since this behavior is not in accord with experiment, we conclude that magnetic anisotropy cannot be due to ordinary dipole-dipole interaction either.

[21]See, e.g., C. Kittel, *Introduction to Solid State Physics*, 3rd ed., Wiley, New York, 1966, p. 375.

If the magnetic moments are not all parallel, then Eq. (4.62) can be generalized to read

$$\mathbf{H} = \sum_i \frac{3(\boldsymbol{\mu}_i \cdot \mathbf{r}_i)\mathbf{r}_i - r^2\boldsymbol{\mu}_i}{4\pi r^5} \qquad (4.63)$$

where $\mathbf{H}$ is the magnetic field acting on a given moment due to all other moments in the sample. For purposes of the following discussion, it is expedient to focus on the interaction between only two of the magnetic moments, $\boldsymbol{\mu}_1$ and $\boldsymbol{\mu}_2$, say. In this case the interaction energy between them is given by

$$V_{\text{od}} = \frac{1}{4\pi r^3}\left[\boldsymbol{\mu}_1 \cdot \boldsymbol{\mu}_2 - \frac{3(\boldsymbol{\mu}_1 \cdot \mathbf{r})(\boldsymbol{\mu}_2 \cdot \mathbf{r})}{r^2}\right] \qquad (4.64)$$

To account for the magnetic crystalline anisotropy, Van Vleck introduced a pseudodipolar interaction identical in form to the ordinary dipole-dipole interaction as far as its magnetic moment variables are concerned but is of much shorter range.[22] With reference to Eq. (4.64) this means that the pseudodipolar interaction is of the form[23]

$$V_{\text{sd}} = f(r)\left[\boldsymbol{\mu}_1 \cdot \boldsymbol{\mu}_2 - \frac{3(\boldsymbol{\mu}_1 \cdot \mathbf{r})(\boldsymbol{\mu}_2 \cdot \mathbf{r})}{r^2}\right] \qquad (4.65)$$

where $f(r)$ is a function that decreases with distance much faster than $1/r^3$. The pseudodipolar interaction is attributed to the combined effects of incomplete quenching of the orbital angular momentum, spin-orbit coupling, and the overlapping of the wave functions of neighboring atoms, as discussed earlier. The strength of this pseudodipolar interaction does account for the right order of magnitude of the observed anisotropy. In addition, the pseudoquadrupolar interaction may be important as well.

There are other forms of anisotropy in ferromagnetic materials. Among these are uniaxial anisotropy induced in magnetic anneal and surface anisotropy due to the lower-order symmetry environment encountered by surface atoms or due to the presence of surface magnetic impurities. Most of these types of anisotropies are important only in thin films; because of a film's very small thickness (~ 1000 Å, say) compared to its other dimensions, the surface atoms may constitute a significant portion of all atoms in a film. We shall now briefly discuss each of these types of anisotropy.

A uniaxial easy axis can be developed in a ferromagnetic crystal by means of magnetic anneal. In this method a static magnetic field is applied in a given direction during the formation of an alloy (e.g., of Ni and Fe) at high temperatures. If the magnetic field $\mathbf{H}_d$ is applied along one of the cubic crystalline easy axes, the uniaxial easy axis induced by magnetic anneal will be aligned in the direction of the annealing field. If $\mathbf{H}_d$ is not applied along a cubic easy axis, the induced easy axis will lie in a direction intermediate between $\mathbf{H}_d$

[22]J. H. Van Vleck, *Phys. Rev.* **52**, 1178 (1937).

[23]Since V_{sd} is quantum-mechanical in origin, the μs and $\mathbf{r}$ should really be operators.

and the nearest cubic easy axis, approaching the direction of $\mathbf{H}_d$ as the magnitude of the annealing field approaches infinity. For a polycrystal the induced easy axis lies in the direction of $\mathbf{H}_d$ since the cubic easy axes are, in this case, randomly oriented.

It appears that the existence of an induced uniaxial easy axis can be attributed mainly to the anisotropic ordering of atom pairs in an alloy. For example, in the case of bulk Permalloy (Ni-Fe), the uniaxial easy axis is induced by the anisotropic diffusion of Fe pairs along and perpendicular to the annealing field; in general, only a few percent more Fe pairs are aligned parallel to $\mathbf{H}_d$ than perpendicular to it. In vacuum or electrodeposited Permalloy films, it is quite likely that imperfection orientation due to magnetostriction also contributes significantly to the existence of the uniaxial anisotropy.

A significant surface anisotropy can exist at the surface of a ferromagnetic film. This type of anisotropy arises because the surface atoms, in contrast to those in the interior, are in an environment of lower symmetry. Owing to this lower-order symmetry, spins of surface atoms tend to orient themselves in certain preferred directions in order to minimize the crystal's free energy.

Because of oxidation, it is quite likely that an antiferromagnetic layer will exist at the surface of a ferromagnetic film. In a Permalloy film, for example, antiferromagnetic materials such as NiO and FeO can exist as surface layers. Since some form of exchange coupling can exist across the ferromagnetic-antiferromagnetic interface, motion of the ferromagnetic spin system may be influenced by the presence of the antiferromagnetic layers. For example, if the ferromagnetic spin system is excited to resonance by an external r.f. field, the ferromagnetic spins near the surface may have a smaller excursion or deviation from their equilibrium directions than those in the interior. Since the antiferromagnetic resonance frequency is in general different from the ferromagnetic one, the antiferromagnetic spins will stay stationary and, by interface exchange, partially "pin down" the ferromagnetic spins near the interface. This phenomenon is very important in the case of standing spin wave resonance in thin films.[24]

Of course, for "spin pinning" to occur, the antiferromagnetic layer must have a finite anisotropy. If this anisotropy is sufficiently large, as in the bulk Co-CoO system, a shift of the hysteresis loop along the H axis to the left also occurs. This type of anisotropy is known as unidirectional anisotropy.[25]

4.6.2 Magnetostatic Energy

The magnetostatic or demagnetizing energy is associated with the magnetic poles existing in or on a ferromagnetic sample. To understand this phenomenon as well as that associated with the Lorentz field introduced earlier, it is necessary to distinguish between the total field and the applied field acting on a given magnetic moment.

[24]R. F. Soohoo, *Magnetic Thin Films*, p. 238.

[25]W. H. Meiklejohn and C. Bean, *Phys. Rev.* **102**, 1413 (1956); **105**, 904 (1957).

Regardless of how the total or local field $\mathbf{H}_{loc}$ acting on a given magnetic moment may be decomposed, it clearly has two distinct origins: one due to the applied field alone and the other due to all the other magnetic moments in the sample. In other words,

$$\mathbf{H}_{loc} = \mathbf{H}_0 + \sum_i \frac{3(\boldsymbol{\mu}_i \cdot \mathbf{r}_i)\mathbf{r}_i - r_i^2 \boldsymbol{\mu}_i}{4\pi r_i^5} \qquad (4.66)$$

where $\mathbf{H}_0$ is the applied field and the summation [which follows from Eq. (4.63)] gives the field acting on a magnetic moment due to all other moments in the sample. Thus, it is reasonable to surmise that $\mathbf{H}_{loc}$ can depend, among other things, on the shape of the sample as well as on the structural symmetry of the crystal.

For mathematical reasons, it is expedient to decompose the second term in Eq. (4.66) into three other terms. Refer now to Fig. 4.9(a), where we first

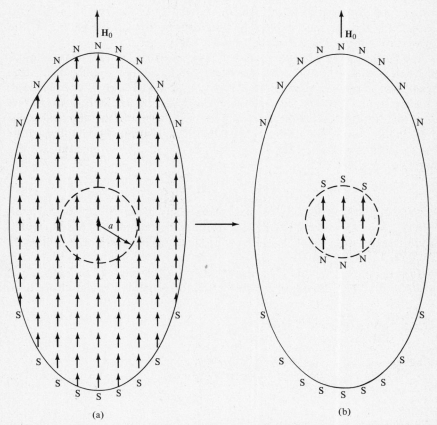

(a)　　　　　　　　　　　　　(b)

Figure 4.9 Decomposition of magnetic field components in a ferromagnet **(a)** showing all magnetic moments and a mathematical sphere and **(b)** with moments outside of the mathematical sphere replaced by magnetic poles on the surfaces of sphere and sample.

draw a mathematical sphere of radius a centered about the magnetic moment under consideration. To calculate the field at this magnetic moment, we clearly can sum up the field components due to all other dipoles in the sample. This summation, however, cannot easily be carried out, because the result depends on the distribution of the dipoles and therefore on the shape of the sample as well as on the symmetry of the crystal. To circumvent this difficulty, we can replace the effect of the dipoles outside the mathematical sphere by a distribution of poles on the surfaces of the sample and sphere, as shown in Fig. 4.9(b). This procedure is permissible provided the sphere is large enough to include a large number of dipoles. In this case the dipoles being replaced are sufficiently far away from the one in question that their discrete distribution is no longer apparent. Then, according to an elementary transformation in electrostatics, the field due to the replaced dipoles is equal to that of the surface pole distributions shown in Fig. 4.9(b). In this context the "surface of the sample" includes the surface of the mathematical sphere as well as the outer surface of the sample. In what follows, we shall calculate the field at the dipole in question due to the pole distribution at the sample's outer surface (known as the demagnetizing field) and the field due to the pole distribution on the surface of the mathematical sphere (known as the Lorentz field).

It was shown by Maxwell that only ellipsoids can be uniformly magnetized when placed in a uniform external magnetic field. If a sufficiently large external field is applied parallel to one of the principal axes of the ellipsoid, the magnetization and the demagnetizing field will be oriented, respectively, parallel and antiparallel to the external field. Furthermore, the demagnetizing field $\mathbf{H}_d$ in this case is uniform in magnitude throughout the body and proportional to the magnetization itself. For this reason, it can be represented by a demagnetizing factor N:

$$\mathbf{H}_d = -N\mathbf{M} \tag{4.67}$$

Since the external field H_0 may be applied along any of the three principal axes, say x, y, z, N in general will have the corresponding values N_x, N_y, N_z. Table 4.2 shows the Ns for a number of ellipsoids of limiting shapes. For any ellipsoid $N_x + N_y + N_z = 1$.

In general, N_x, N_y, and N_z are dependent on the ratios of the axes of the ellipsoid a, b, c. The demagnetization factors for general ellipsoids have been

TABLE 4.2   DEMAGNETIZING FACTORS FOR
              LIMITING ELLIPSOIDAL SHAPES

Shape	Axis	N
Sphere	All three	$\frac{1}{3}$
Thin film	Normal	1
Thin film	In plane	0
Long circular cylinder	Longitudinal	0
Long circular cylinder	Transverse	$\frac{1}{2}$

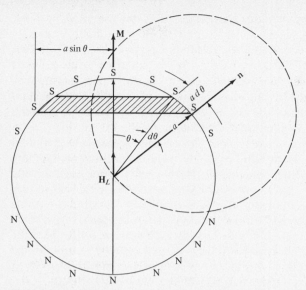

Figure 4.10   Fictitious sphere in Lorentz field calculation.

calculated by Osborn.[26] For nonellipsoids, the demagnetizing field is in general nonuniform in distribution. However, if the external field is large enough to ensure magnetic saturation, it can be shown that the sum of N_x, N_y, and N_z is still unity, as for the case of an ellipsoid.[27]

The Lorentz field $\mathbf{H}_L$ due to the magnetic poles on the surface of the mathematical (fictitious) sphere can be calculated as follows. Refer to Fig. 4.10, where the surface pole density for an elemental area oriented in a direction making an angle θ with the magnetization $\mathbf{M}$ is seen to be simply $-M \cos \theta$; the minus sign follows directly from the pole distribution depicted in Fig. 4.10. Correspondingly, $(\hat{n}M \cos \theta)/4\pi a^2$ is the field per unit area produced at the center of the sphere by the pole density $-M \cos \theta$. This is because the pole density on a surface element at angle θ and residing on the plane of the paper gives rise to a spherically symmetric magnetic field and is oriented in the direction θ at the center of the cavity. The magnitude of this field per unit area is just $(M \cos \theta)/4\pi a^2$, as is evident from a consideration of the field per unit area on the dotted spherical surface shown in Fig. 4.10. From the field distribution plot of Fig. 4.10 obtained from a consideration of symmetry, we see that the total field $\mathbf{H}_L$ at the cavity center due to all the pole density elements on the surface of the cavity should be oriented in the direction of $\mathbf{M}$. For this reason, to obtain the Lorentz field $\mathbf{H}_L$ at the center of the cavity, we should integrate the projection of $(\hat{n}M \cos \theta)/4\pi a^2$ along $\mathbf{M}$, or $[(\mathbf{M} \cos \theta)/4\pi a^2] \cos \theta$,

[26]J. A. Osborn, *Phys. Rev.* **67**, 351 (1945).

[27]E. Schlömann, *J. Appl. Phys.* **33**, 2825 (1962).

over the entire surface of the fictitious cavity:

$$\mathbf{H}_L = \int_0^\pi \frac{\mathbf{M}\cos\theta}{4\pi a^2}\cos\theta\,(2\pi a\sin\theta)(a\,d\theta) = \frac{\mathbf{M}}{3} \tag{4.68}$$

Since the location of the cavity center is arbitrary, the value obtained for the Lorentz field $\mathbf{H}_L$ is applicable to every magnetic moment in a magnetized specimen. Thus, we conclude that *in any magnetized medium* there is a Lorentz field of magnitude $M/3$ in the direction of $\mathbf{M}$ acting to align it.

To complete our analysis of the field components within an ellipsoidal sample placed in a saturation uniform magnetic field, we must now compute the field H_c due to the magnetic moments within the cavity. Since the magnetic moments in the cavity are not necessarily far removed from the one under consideration, their exact locations or, equivalently, the crystal structure of the material must be considered. Since most ferromagnetic materials of practical interest are cubic in nature, we shall examine this case in particular. Since the applied field is assumed to be sufficiently large as to saturate the sample, the magnetic moments within the cavity are parallel to each other and to $\mathbf{H}_0$. It follows from Eq. (4.66) that at the location of the magnetic moment under consideration, the z component of the magnetic field caused by the other moments $\boldsymbol{\mu}_i$ is

$$H_{cz} = \sum_i \frac{3\mu_i z_i^2 - \mu_i r_i^2}{r_i^5} \tag{4.69}$$

where z_i is the projection of $\mathbf{r}_i$ along the z or H_0 axis. By the symmetry of lattice and cavity, the same argument in connection with Eq. (4.62) applies, and it again follows that

$$\sum_i \frac{r_i^2}{r_i^5} = \sum_i \frac{x_i^2 + y_i^2 + z_i^2}{r_i^5}$$

is equal to $3\sum_i z_i^2/r_i^5$. Accordingly, H_{cz} of Eq. (4.69) is identically zero. Furthermore, since the $\boldsymbol{\mu}_i$ are all aligned in the z direction, it follows from Eq. (4.66) that H_{cx} and H_{cy} are also zero.

For crystal with other than cubic symmetry, $\mathbf{H}_c$ is in general nonzero. Values of $\mathbf{H}_c$ for tetragonal and simple hexagonal lattices have been given by Mueller.[28]

Let us now summarize the results obtained in this section before applying them to the problem of domain formation. Briefly, we have found that if an ellipsoidal sample is placed in a uniform magnetic field $\mathbf{H}_0$ applied along one of its principal axes, the total or local field inside the sample is given by

$$\mathbf{H}_{loc} = \mathbf{H}_0 + \mathbf{H}_D + \mathbf{H}_L + \mathbf{H}_c \tag{4.70}$$

where $\mathbf{H}_D = -N\mathbf{M}$ is the demagnetizing field, $\mathbf{H}_L = \mathbf{M}/3$ the Lorentz field, and $\mathbf{H}_c$ the only field component whose value is dependent upon the crystal struc-

[28]H. Mueller, *Phys. Rev.* **47**, 947 (1935); **50**, 547 (1936).

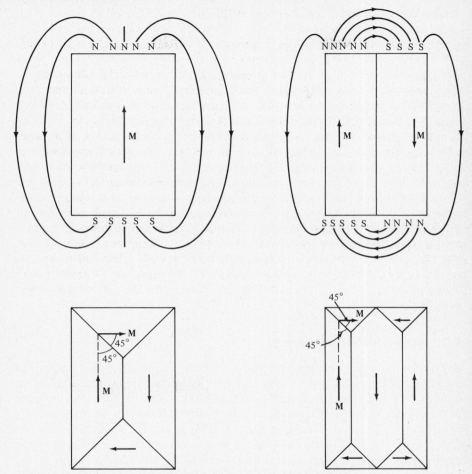

Figure 4.11 Domain configuration for minimum demagnetization energy.

ture. Of these four component fields, H_c and H_L are fields acting on a given magnetic moment inside a fictitious cavity due, respectively, to other moments inside the cavity and magnetic poles on the walls of the cavity. As such, they, like the exchange field, tend to align the moments parallel. They need not, however, be explicitly considered in the problem of domain-wall formation or dynamics. For this reason, we need only think of minimizing the demagnetizing energy associated with H_D in constructing ferromagnetic domains. For an ellipsoid magnetized along one of its principal axes, this energy is simply equal to $\frac{1}{2}NM^2$ where N is the demagnetizing factor along the axis in question.

Refer now to Fig. 4.11(a), where we have sketched the external field lines due to the surface magnetic poles.[29] For this purpose, we can think of the

[29]For simplicity, we consider here a two-dimensional case. Although the figure shown is not an ellipsoid, our conclusions based on it are conceptually valid.

demagnetizing energy as being stored in the surrounding field. This energy can be reduced by reducing the spatial extent of this storage, as shown in Fig. 4.11(b), where the magnetization in half of the sample has been reversed in direction.

If the crystal under consideration is cubic with two of its easy axes lying along the vertical and horizontal directions, the demagnetizing energy can be reduced to zero without changing the anisotropy energy by dividing up the crystal into two horizontal and two vertical domains as shown in Fig. 4.11(c). Since **M** in any of the four domains is at 45° to the nonvertical intervening boundary surfaces, the normal component of **M**, being equal to $M/\sqrt{2}$ on either side of the boundary, is continuous, alleviating the appearance of any magnetic poles [see boundary condition (2.46)]. For such a closed magnetic circuit with no surrounding field, the demagnetizing energy clearly vanishes.

If anisotropy and magnetostatic energies were the only energy components of consequence in domain formation, the domain configurations of Fig. 4.11(c) and 4.11(d) should be equally probable since in either case the anisotropy and magnetostatic energies are minimized. This is not really the case, however, because there are energy components related to the dependence of crystal strain on **M** and vice versa (magnetostriction) and to the domain wall itself that must also be taken into account. These new energy components will now be discussed.

4.6.3 Magnetostrictive Energy

Since the equilibrium magnetization direction in a crystal is influenced by the structure of the crystal, it is not unreasonable to surmise that mechanical forces applied to a solid can, in some measure at least, influence the magnetic state of the crystal. Conversely, a change in the magnetization may change the state of strain of the crystal. The latter phenomenon is referred to as *magnetostriction*. Physically, it is useful to think of magnetostriction as arising from the dependence of the crystalline anisotropy energy on the state of strain of the lattice. Thus, it may be energetically favorable for the crystal to deform slightly from the exact cubic condition if doing so will lower the anisotropy energy more than it raises the elastic energy.

For isotropic materials it is easy to relate the change in length of the specimen to the angle θ between **M** and the direction in which δl is measured. To begin with, we would expect δl to have the same value for $\theta = 0$ and π. The simplest function with this property is of the form $\cos^2 \theta$. Now, for a demagnetized specimen the magnetization vectors of the domains are randomly oriented. In this case we would therefore expect δl measured in any given direction to be zero. Since $\cos^2 \theta$ averaged over all solid angles is $\frac{1}{3}$, we must have

$$\frac{\delta l}{l} = A(\cos^2 \theta - \tfrac{1}{3}) \tag{4.71}$$

where l is the length of the specimen in the direction of measurement and A is a constant. At saturation, the direction of **M** coincides with the direction in which δl is measured; i.e., $\theta = 0$. If we designate $(\delta l/l)_{\theta=0}$ as the saturation magnetostriction λ_s, then A in Eq. (4.71) can be evaluated to yield $3\lambda_s/2$. Thus, Eq. (4.71) takes the form

$$\frac{\delta l}{l} = \tfrac{3}{2}\lambda_s(\cos^2\theta - \tfrac{1}{3}) \tag{4.72}$$

Since this equation is derived for an isotropic material, it is useful for the interpretation of results of magnetostriction experiments involving polycrystalline samples. Within such samples, the crystalline axes are assumed to be randomly oriented so that the concept of a single magnetostriction value is plausible.

For a single crystal the corresponding expression for $\delta l/l$ is much more complicated. If the crystal is cubic, detailed calculations show that[30]

$$\frac{\delta l}{l} = \tfrac{3}{2}\,\lambda_{[100]}(\alpha_1^2\beta_1^2 + \alpha_2^2\beta_2^2 + \alpha_3^2\beta_3^2 - \tfrac{1}{3})$$

$$+ 3\lambda_{[111]}(\alpha_1\alpha_2\beta_1\beta_2 + \alpha_2\alpha_3\beta_2\beta_3 + \alpha_3\alpha_1\beta_3\beta_1) \tag{4.73}$$

where α_1, α_2, α_3 are direction cosines of $\mathbf{M}$ and β_1, β_2, β_3 are direction cosines of the measuring direction. Note that if the saturation magnetostrictions along the [100] and [111] directions, i.e., $\lambda_{[100]}$ and $\lambda_{[111]}$, are assumed equal to λ_s, Eq. (4.72) results wherein θ is defined as equal to $\cos^{-1}\sum_i \alpha_i\beta_i$. The expression $\delta l/l$ given by Eq. (4.73) is adequate for the description of the experimental data for most cubic crystals. In some cases higher-order terms must be considered.

The saturation magnetostriction measured along various crystalline axes of different materials can be either positive or negative signifying, respectively, elongation or contraction in the direction of magnetization. For iron $\lambda_{[100]}$, $\lambda_{[111]}$, and λ_s are, respectively, 19.6×10^{-6}, 18.8×10^{-6}, and -7×10^{-6} at room temperature. For nickel the corresponding values are -46×10^{-6}, -25×10^{-6}, and -34×10^{-6}. Because of the smallness of the fractional change in length represented by these values of λ, rather sensitive instruments are necessary for their measurement.

From the above data we conclude, for example, that the length of a crystal along the cubic edge of an iron single crystal is extended by $\lambda_{[100]}$ or about 20 ppm when the crystal is magnetized in this direction. At the same time there is a contraction of about $\lambda_{[100]}/2$ in the direction perpendicular to $\mathbf{M}$. Consequently, when the upper and lower domains form in Fig. 4.11(c), the lattice structure in these domains extends in the magnetization direction, if free to do so, by an amount $3\lambda_s/2$. If a mechanical restraint is placed on this movement, as in a solid, a component of magnetostrictive energy is introduced. For this reason alone, the free energies associated with the domain configurations of Fig. 4.11(c) and 4.11(d) are not equal and they do not occur with equal probability.

If the volume of the strain-producing horizontal domains is reduced, by going from state 4.11(c) to state 4.11(d), for example, this component of energy

[30]A. H. Morrish, *The Physical Principles of Magnetism*, Wiley, New York, 1965, p. 324.

will also be reduced. As more vertical and horizontal domains are introduced with **M** in adjacent vertical and horizontal domains making a constant angle of 45° with the intervening surface, the horizontal or edge domains become wedge-shaped. Their total volume becomes smaller as the vertical domain boundary surfaces come closer together, thus forming thin plane domains like pages of a book. As this lamination process continues, however, the number of intervening surfaces called domain walls also increases. The energy associated with these walls (see Section 4.6.4) thus must also be considered to determine the minimum energy domain configuration.

4.6.4 Domain Wall Energy

Consider the exchange energy between two neighboring spins. If the angle between these spins, each of magnitude S, is θ, Eq. (4.53) becomes:

$$E_{ex} = -2JS^2 \cos \theta \qquad (4.74)$$

where J is the exchange integral between the spins. Thus, the energy difference between a system of two exchange-coupled parallel spins and that of two antiparallel spins is simply $4JS^2$. On the other hand, if θ is very small, the increase in exchange energy compared to that of the parallel state is $JS^2\theta^2$. Thus, the increase in exchange energy due to the presence of domain boundaries depicted in Fig. 4.11 will depend on whether the spin rotation is accomplished in one or many steps.

From Eqs. (4.53) and (4.74) we can easily show that the ratio of the increase in exchange energy for the distributed and discrete cases is given by

$$\frac{(\Delta E_{ex})_{dist}}{(\Delta E_{ex})_{disc}} = \frac{\theta_0^2}{2(N-1)(1-\cos \theta_0)} \qquad (4.75)$$

where θ_0 is the total change in spin angle across the domain boundary and N is the number of spins participating in the rotation.

From Eq. (4.75) we see that for a given value of θ_0, e.g., π or $\pi/2$, $(\Delta E_{ex})_{dist}/(\Delta E_{ex})_{disc}$ is nearly inversely proportional to N for $N \gg 1$. If domain wall energy were the only component of energy of consequence in determining the width of the wall, N and therefore the wall size would increase without limit in order to minimize the energy of the sample. In practice, this limitless expansion does not occur since other components of the free energy, e.g., the anisotropy energy, come into play to arrest this growth. Suppose that the spins within the adjacent domains are aligned in the direction of the anisotropy easy axes; the spins within the intervening domain wall then in general point away from the easy axes. Therefore, as the wall expands in size, the anisotropy energy also increases, and this increase tends to compensate for the decrease in exchange energy as N increases. For this reason, the equilibrium domain wall width is finite and is on the order of a few hundred angstroms in ordinary ferromagnetic materials. There are two kinds of common domain walls, one in which the spins rotate about an axis perpendicular to the plane of the wall (called the *Bloch wall*) and one in which the spins rotate about an axis in the plane of the wall itself (called the *Néel wall*). These cases are illustrated in Fig. 4.12. In practice, there may

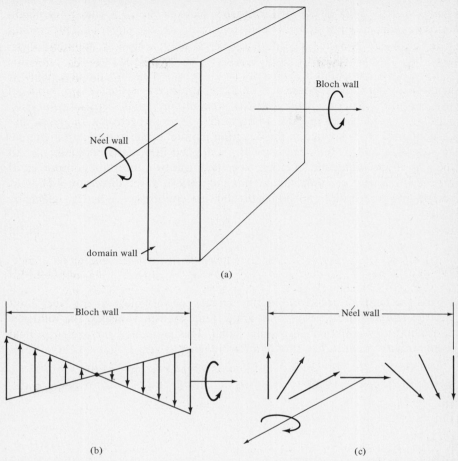

Figure 4.12 **(a)** Direction of spin rotation in Bloch and Néel walls, **(b)** Spin projection in a 180° Bloch wall and **(c)** spin orientation in a 180° Néel wall.

be domain walls of more complex configuration than those depicted, the exact shapes of domains and domain walls being dependent on the delicate balance of different components of energy studied in this section. Furthermore, the presence of imperfections may drastically modify the shape as well as the location of the walls in a ferromagnet.

4.7 ANTIFERROMAGNETISM

In connection with Eq. (4.53), $E_{ex} = -2J_{ij}S_i \cdot S_j$ we mentioned that if J_{ij} is positive, ferromagnetism results, while if J_{ij} is negative, antiferromagnetism results. In the former case E_{ex} attains its negative maximum value if S_i is parallel to S_j, whereas in the latter case it attains its negative maximum value if S_i is antiparallel to S_j. The sign of J_{ij} is dependent on the degree of overlap of the charge distributions of the ferromagnetic electrons. Since S_i and S_j of neighboring atoms are oppositely directed, antiferromagnetic materials that contain many such subassemblies have no net magnetization. This does not imply, however, that the presence of such magnetic states cannot be detected. Since the antiparallel alignment of the spins can be disturbed by application of a magnetic field, these substances possess a finite susceptibility. Usually, two kinds of susceptibility are considered: parallel susceptibility and perpendicular susceptibility, as depicted in Figs. 4.13(a) and 4.13(b), respectively. In the former case the field H_0 is applied in the direction parallel to the axis of antialignment of μ_i and μ_j, and in the latter case H_0 is applied in a plane perpendicular to the axis of antialignment. In either case there will be a resultant component of magnetic moment per unit volume (magnetization) in the direction of H_0 corresponding to parallel and perpendicular susceptibilities $\chi_{\parallel}$ and $\chi_{\perp}$ given by

$$\chi_{\parallel} = \frac{\Delta M_{\parallel}}{H_0} \tag{4.76}$$

and

$$\chi_{\parallel} = \frac{\Delta M_{\perp}}{H_0} \tag{4.77}$$

Unless the temperature of the antiferromagnet is very close to the Néel point T_N, the temperature above which antiferromagnetism vanishes, the influence of a laboratory-applied field is quite small. This is because the exchange field in an antiferromagnet, like that in a ferromagnet, is very large ($\sim 10^6$ Oe) com-

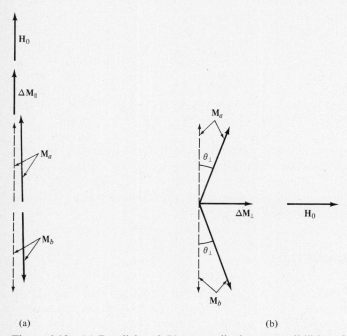

(a)   (b)

Figure 4.13 (a) Parallel and (b) perpendicular susceptibilities of an antiferromagnet.

pared to any field available in a laboratory. Thus, $\Delta\mathbf{M}_{||}$, $\Delta\mathbf{M}_{\perp}$, and therefore $\chi_{||}$ and $\chi_{\perp}$ are in general all very small.

Because of their low static susceptibility, the use of antiferromagnetic materials in practical devices is very limited. However, a lightweight resonance isolator at millimeter wave frequencies can be constructed using an antiferromagnetic material if the frequency corresponding to the exchange field is set equal to the operating frequency.

4.8 FERRIMAGNETISM

Ferromagnetic materials have a common disadvantage at microwave frequencies. They are in general metallic in nature, and their low resistivity gives rise to intolerably high eddy current losses at microwave frequencies; for ordinary ferromagnetic materials the eddy current power loss is proportional to the square of the frequency.[31] In principle, at least, these losses, being proportional to the cube of sample thickness, can be reduced by laminating the ferromagnetic sample to restrict eddy current paths. However, there is a practical limit to the thinness to which discrete laminations can be satisfactorily manufactured. In fact, even for vacuum-deposited thin ferromagnetic films (~ 100–$10,000$ Å), the eddy current losses are still too large to be usable at microwave frequencies. The only alternative to this dilemma is to use magnetic materials of unusually high resistivity since the eddy current power loss is inversely proportional to the resistivity. Ferrimagnetic materials, known as ferrites and garnets, having a resistivity of up to 10^7 Ωm as compared to 10^{-7} Ωm for iron, reduce eddy current losses in them to negligible values, even at microwave and millimeter wave frequencies.

Ferrites are mixed metallic oxides of high resistivity with a ceramic appearance. They are made by sintering a mixture of metallic oxides of appropriate proportions. Although iron ferrite or magnetite was known in ancient times, as lodestone, the first usable modern ferrite was made only in the early 1940s.[32] These materials found immediate use in carrier, radio, television, and later in computer core-memory applications. At the end of 1953 the first commercially available ferrite devices, a ferrite-loaded waveguide with different attenuation in the two opposite directions of propagation, appeared on the market.

Conceptually, we can think of ferrimagnetism as uncompensated antiferromagnetism. As discussed below, there are two sublattices in ferrimagnetic materials and the magnetic moments in these two sublattices are oppositely aligned. Unlike the case of antiferromagnetism, however, the moments of the two sublattices are not equal in magnitude. Consequently, there is a net magnetic moment for the sample as a whole. Owing to its "pseudo-antiferromagnetic" nature, ferrimagnetic materials usually have a lower magnetization than

[31]R. F. Soohoo, *Magnetic Thin Films*, p. 138.

[32]J. L. Snoek, *Philips Tech. Rev.* **8**, 353 (1946).

ferromagnetic materials. For this reason, they are not used in such applications as high-flux-density low-frequency core materials for power transformers. However, as far as high-frequency applications are concerned, they have the distinct advantage of having extremely high resistivity, thus minimizing eddy current losses. This latter property arises mainly because, in contrast to the case of ferromagnets, the constituent ions, both oxygen and metallic, have no free $3s$ or $4s$ electrons available for conduction. Whereas the wave functions of ferromagnetic electrons directly overlap, giving rise to exchange coupling, exchange coupling between electron spins at the sublattices is indirect in nature. Since the metallic ions are located at the interstices formed by the oxygen ions, they are too far apart for direct wave function overlap to occur. Instead, indirect exchange interaction occurs with the intervening oxygen ion as the intermediary. The wave function of the metallic ion at sublattice A overlaps with that of the oxygen ion and the wave function of the oxygen ion in turn overlaps with that of the metallic ion at sublattice B. It turns out that the interaction between ions at the A sites as well as the interaction between ions at the B sites are both antiferromagnetic. However, since the A-B interaction between the ions at A and B sites is overridingly strong and antiferromagnetic in nature, ferrimagnetism results provided the magnetic moments at sublattices A and B are not equal in magnitude.[33]

To proceed beyond our general discussion of ferrimagnetism, we must now refer to the ionic and crystal structure of these materials. Ferrites are ceramic ferrimagnetic materials with the general chemical composition $MO \cdot Fe_2O_3$ where M is a divalent metal such as iron, manganese, magnesium, nickel, zinc, cadmium, cobalt, copper, or a mixture of these. The ferrites crystallize into the spinel structure, which is named after the mineral spinel, $MgAl_2O_4$. Because the radius of the oxygen ion is several times larger than the radii of the metallic ions (see Table 4.3), the spinel crystal structure is primarily determined by the positioning of the oxygen ions. Indeed, the crystal structure can be thought of as being made up of nearly the closest possible packing of layers of oxygen ions, with the metallic ions fitted in at the interstices.

This ion-packing situation is depicted in Fig. 4.14(a). The bottom layer of oxygen ions is represented by dotted circles centered at points marked with solid dots. The second layer, if drawn, would constitute circles centered at points marked with crosses. Similarly, the third layer, if drawn, would consist of circles centered at points marked with small circles. This arrangement of ions gives rise to the cubic close-packed or face-centered cubic lattice; Fig. 4.14(a) depicts the arrangement of atoms in the [111] crystal plane. Actually, the oxygen ions in ferrites are somewhat further apart than depicted in Fig. 4.14. Nevertheless, this figure is very helpful in giving a physical picture of the existence of two possible sites A and B for the metallic ions. In Fig. 4.14(b) we have superimposed the second-layer ions (solid circles) on top of those of the first layer (dotted circles). On close examination, it can be seen that there are two kinds of interstices, denoted A and B. A metallic ion located at the A site has

[33]L. Néel, *Ann. Phys.* **3**, 137 (1948).

TABLE 4.3   IONIC RADII OF
SEVERAL ELEMENTS[a]

Ion	Radius (Å)	
O^{2-}	1.32^a	1.40^b
Fe^{3+}	0.67	—
Fe^{2+}	0.83	0.75
Mn^{2+}	0.91	0.80
Mg^{2+}	0.78	0.65
Ni^{2+}	0.78	0.69
Zn^{2+}	0.83	0.74
Cd^{2+}	1.03	0.97
Co^{2+}	0.82	0.72
Cu^{2+}	0.70	—

Source: Charles D. Hodgman (ed.),
Handbook of Chemistry and Physics, 36th
ed., Chemical Rubble Co., Cleveland,
Ohio, 1954–1955, pp. 3095–3097.

[a] Computed by S. Goldschmidt on the
basis of empirical assumptions.

[b] Computed by L. Pauling on the basis of
wave mechanics.

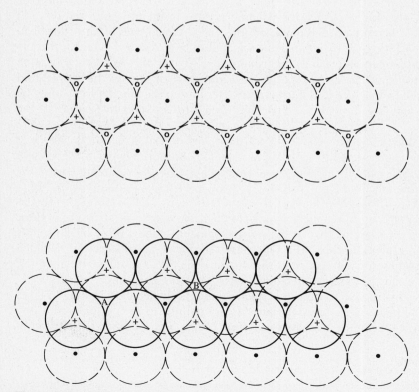

Figure 4.14   Cubic close packed or face centered cubic lattice.

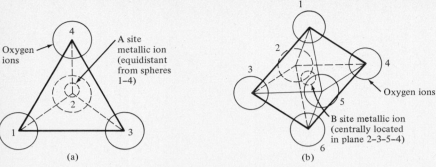

Figure 4.15 **(a)** Tetrahedral coordination: Spheres 1–3 correspond to top-layer oxygen ions, sphere 4 to a bottom-layer oxygen ion in Fig. 4.14(b). **(b)** Octahedral coordination: Spheres 1–3 correspond to top-layer oxygen ions, and spheres 4–6 correspond to bottom-layer oxygen ions in Fig. 4.14(b).

four nearest oxygen ion neighbors, three oxygen ions in the top layer and one in the bottom layer. (There are also equivalent sites with three oxygen ions in the bottom layer and one at the top layer.) In other words, it is in a site of tetrahedral coordination [see Fig. 4.15(a)]. If the metallic ion is located at site B, it will have six nearest oxygen ion neighbors, three in the top layer and three in the bottom layer. Since these oxygen ions are at the corners of an octahedron [see Fig. 4.15(b)], the metallic ion at B is said to be in a site of octahedral coordination. Similar conclusions regarding the existence of A and B sites can be drawn from our examination of the superposition of ions in the second and third layer, in the third and fourth layers, etc.; ions in the fourth layer should be situated just like those in the first layer owing to the repetitive nature of the crystal.

A diagram of the unit cell[34] of the spinel crystal structure is shown in Fig. 4.16 where the Mg^{2+}, Al^{3+} and O_4^{2-} ions are represented by spheres of different sizes (entirely arbitrary) and shades. For zinc and cadmium ferrites the divalent metallic ions Zn^{2+} or Cd^{2+} are at the A sites (sites occupied by Mg^{2+} ions in Fig. 4.16), while the two trivalent ferric ions $2Fe^{3+}$ are at the B sites (sites occupied by Al^{3+} ions in Fig. 4.16). In other words, the structure of these ferrites is the same as that of the normal spinel. Actually, most of the simple ferrites [e.g, nickel ferrite ($NiFe_2O_3$)] have the inverse spinel structure, in which one trivalent ferric ion Fe^{3+} is at the A sites while the remaining trivalent ferric ion Fe^{3+} and the divalent metallic ion M^{2+} are at the B sites.[35]

It is instructive to note that each atomic layer depicted in Fig. 4.14 lies in a [111] plane. These crystal planes are all parallel to each other with a perpendicular spacing of $a/\sqrt{3}$ between them where a is the linear dimension of

[34]A unit cell is a basic pattern unit composed of various atoms in well-defined positions; if it is repeated regularly in all three dimensions, a crystal results.

[35]This can be confirmed by neutron diffraction experiments. See C. G. Shull, E. O. Wollan, and W. C. Koehler, *Phys. Rev.* **84**, 912 (1951).

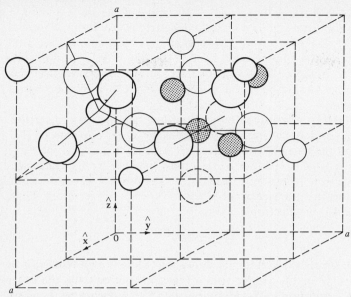

Figure 4.16 Unit cell of spinel structure. The position of the ions in only two octants is shown. The dashed circles belong to other octants. The solid lines indicate the four- and sixfold coordination of the respective metal-ion positions. Large circles: oxygen ions; small shaded circles: metal ions at octahedral sites; small unshaded circles; metal ions at tetrahedral sites. [After E. W. Gorter, *Phillips Res. Rep.* 9, 295–320 (1954).]

the unit cell; note that a plane passing through the points $\hat{x}a$, $\hat{y}a$, $\hat{z}a$ is one such [111] plane. It is left as an exercise for the reader to associate the ions shown in Fig. 4.16 with those shown in Figs. 4.14 and 4.15.

We have mentioned previously that the magnetic moments at the A sites are antiparallel to those at the B sites. Consider specifically the case of iron ferrites, Fe_3O_4. In ionic form Fe_3O_4 can be represented by the formula

$$Fe_3O_4 = FeO \cdot Fe_2O_3 = Fe^{3+}(Fe^{3+}Fe^{2+})O_4^{2-} \qquad (4.78)$$

As indicated, one Fe^{3+} ion is at the A site while one Fe^{3+} ion and one Fe^{2+} ion are at the B site. According to Eq. (4.78), for every formula unit of Fe_3O_4, there are two Fe^{3+} ions and one Fe^{2+} ion. According to Table 4.1, Fe^{3+} and Fe^{2+} have the following electronic configurations:

$$Fe^{3+}: \qquad (1s^2 2s^2 2p^6 3s^2 3p^6)3d^5 \qquad (4.79)$$

$$Fe^{2+}: \qquad (1s^2 2s^2 2p^6 3s^2 3p^6)3d^6 \qquad (4.80)$$

Hund's rule states in part that the electron spins of the $3d$ electrons add in such a way as to give the maximum possible S consistent with the Pauli exclusion principle. It follows that Fe^{3+} has a magnetic moment of $5\mu_B$ while Fe^{2+} has a magnetic moment of $4\mu_B$, since the $3d$ shell can accommodate only a maximum of ten electrons and the net spin of a close shell must be zero. Thus, if the spin of the Fe^{3+} ion at the A site were parallel to the spins of the Fe^{3+} and Fe^{2+} ions at the B site, the total number of Bohr magnetons per formula unit of Fe_3O_4 would be $5 \times 2 + 4 \times 1 = 14$. However, this is at great variance with the experimental value of 4. On the other hand, if the spin of Fe^{3+} at the A site is assumed to be antiparallel to those of Fe^{3+} and Fe^{2+} at the B sites, the effective magneton number should be $(5 \times 1 + 4 \times 1) - 5 \times 1 = 4$, in good agreement with experiment.

Let us now turn to a consideration of the low conductivity in ferrites. Specifically, again consider the case of iron ferrite, Fe_3O_4. According to Eqs. (4.78)–(4.80), each of the two Fe^{3+} ions gives up three electrons, and the Fe^{2+} ion gives up two electrons to the four oxygen ions. In other words, eight electrons from the three iron atoms are transferred to the four oxygen atoms in a formula unit, resulting in a transfer of two electrons to each oxygen atom. Thus, according to Table 4.1, the electronic configuration in O^{2-} is

$$O^{2-}: \quad (1s^2 2s^2 2p^6) \tag{4.81}$$

An examination of formulas (4.79)–(4.81) shows that there are no free $3s$ or $4s$ electrons in the crystal Fe_3O_4 to give rise to high conductivity. On the other hand, metallic iron, according to Table 4.1, has two free $4s$ electrons that contribute to conduction in this ferrimagnetic metal.

It is interesting to note from Eq. (4.78) that if an electron were to be transferred from the divalent ferrous ion Fe^{2+} at a B site to the trivalent ferric ion Fe^{3+} also at a B site, the ionic form of the crystal would remain completely unchanged. Thus, it might be expected that electron flow can be facilitated via this mechanism, giving rise to significant conductivity. Indeed, it turns out that the resistivity of iron ferrite is about $10^{-4}\,\Omega\,m$, which though three orders of magnitude higher than that of iron is still too low for high-frequency applications. However, if the ferrous ion Fe^{2+} at the B site is replaced by one of the divalent metallic ions listed in Table 4.3 (or an admixture thereof), the resistivity can be greatly increased to a value as high as $10^7\,\Omega\,m$. For example, if Fe^{2+} is replaced by Ni^{2+}, chemical formula (4.78) becomes

$$NiO \cdot Fe_2O_3 = Fe^{3+}(Fe^{3+}Ni^{2+})O_4^{2-} \tag{4.82}$$

It is now no longer possible for electrons to be interchanged between the trivalent Fe^{3+} ion and divalent Ni^{2+} ion, although both are at the B sites, without producing a final state different from the initial one. Since all obvious mechanisms of conduction have been eliminated, the resistivity of these ferrites should be very high. Based on this discussion, it is helpful to think of ferrites as being derived from magnetite or iron ferrite, Fe_3O_4, through the replacement of ferrous ions by appropriate divalent ions.

There are notable exceptions to the $MO \cdot Fe_2O_3$ formula for ferrites discussed above. The most important class of materials in this category are the rare-earth garnets. They have the general chemical formula $5Fe_2O_3 \cdot 3M_2O_3$ where M represents yttrium or some other rare-earth ion from samarium to

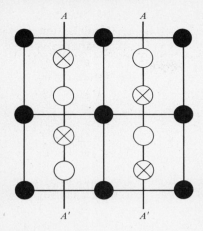

Figure 4.17 Two-dimensional view of a garnet unit cell. (Filled circles: a sites; crossed circles: c sites; open circles: d sites.)

lutecium in the periodic table.[36,37] This structure differs from the spinel lattice of the conventional ferrites in several aspects. Two of the chief structural differences are that the garnet has three types of lattice sites available to metallic ions as compared to the two types of sites in the spinel and that all of the possible sites in the garnet lattice are filled, as opposed to the half-filled B sites and the one-eighth-filled A sites of the spinel lattice. The arrangement of atoms in a unit cell of a rare-earth garnet is shown in Fig. 4.17. Chiefly because all sites in a garnet are filled, its single crystal linewidth is extremely narrow, in the order of an oersted or less. Their small linewidths render garnets very useful in nonlinear and low microwave frequency devices as well as in microwave filter applications.

Another exception to the $MO \cdot Fe_2O_3$ formula for ordinary ferrites is Ferroxdure,[38] a material used for permanent magnets, with a chemical formula $BaO \cdot 6Fe_2O_3$. It has a hexagonal structure rather than the cubic crystal structure of the spinel, giving rise to a high uniaxial anisotropy. Because of its high anisotropy field, it self-resonates at about 50 GHz with no externally applied field. Consequently, this material can be used to build light and compact resonance isolators at these millimeter wave frequencies.

As mentioned previously, polycrystalline ferrites are ceramic materials made by sintering various ferromagnetic oxides at high temperatures. The preparation of these ferrites requires a great deal of art because the optimum preparation condition cannot easily be deduced from first principles. For this reason, the exact composition, firing temperature, and furnace atmosphere required to produce ferrites with a particular set of characteristics are determined by repeated experimentation. A discussion of the preparation of these materials is given in Section 8.7.

[36]F. Bertaut and F. Fomat, *C. R. Acad. Sci.* **242**, 382 (1956).

[37]G. S. Geller and M. A. Gilleo, *Acta Crystallogr.* **10**, 239 (1957).

[38]J. J. Went, G. W. Ratheman, E. W. Gorter, and G. Oosterhout, *Philips Tech. Rev.* **13**, 194 (1952).

PROBLEMS

4.1 Using Hund's rule, find S, L, and J for Cr^{2+}, Gd^{3+}, and Er^{3+}.

4.2 Derive Eq. (4.31) for the Landé g-factor.

4.3 Many paramagnets obey the Curie-Weiss law of Eq. (4.46) with small values of T_c on the order of $10°K$ or less rather than the Curie law of Eq. (4.41). Consider a sample composed of such a paramagnetic film of thickness δ coated on top of a diamagnetic plane substrate of area A, thickness t, and susceptibility χ_d. Find the temperature at which the sample as a whole is nonmagnetic.

4.4 Figure 4.18 shows an ellipsoid sample with uniaxial-induced anisotropy $e_k = K_1 \sin^2 \theta$ where θ is the angle between **M** and the easy axis. If **M** is initially oriented in the $+z$ direction, find the coercive force H_c if the switching field $\mathbf{H}_0$ is applied in the $-z$ direction. H_c is defined as the values of H_0 required to reduce the **M** component along $\mathbf{H}_0$ to zero. Assume that the demagnetizing factors N_x, N_y, and N_z are given.

4.5 Figure 4.19 shows the domain configuration after saturation in the hard direction and with the subsequent reduction to zero field for a Permalloy film. Given that the anisotropy energy density $e_k = K_1 \sin^2 \theta$ where θ is the angle that **M** makes with the easy axis and that σ_w is equal to the wall energy per unit wall surface area, find the expression for the equilibrium value of D.

4.6 Referring to Fig. 4.13 and using the torque equation $\tau = \mu_0 \mu_i \times \mathbf{H}_0$, show how an application of $\mathbf{H}_0$ can result in $\Delta\mathbf{M}_\perp$. How does $\Delta\mathbf{M}_\parallel$ arise?

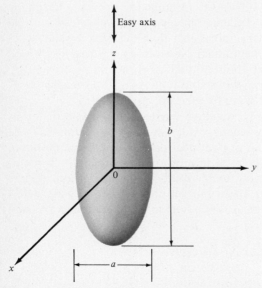

Figure 4.18 Ellipsoidal particle with uniaxial anisotropy and $b > a$.

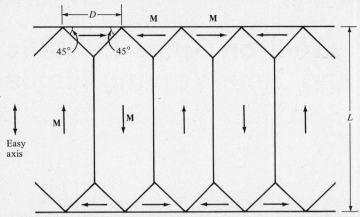

Figure 4.19 Equilibrium domain configuration in a Permalloy film of thickness t.

4.7 (a) Calculate the saturation magnetization at $0°K$ in ampere turns per meter for the following ferrites: $MnO \cdot Fe_2O_3$, $FeO \cdot Fe_2O_3$, and $CdO \cdot Fe_2O_3$.

(b) If the mole fraction of zinc ferrite in a nickel-zinc ferrite sample is equal to δ, write down the ionic formula for the nickel-zinc ferrite. Indicate which ions are at the B site.

Chapter 5

Electron Spins in Static and Time-Varying Fields

In this chapter we shall examine the behavior of electron spins, singly or in aggregate, in the presence of static and r.f. fields. Concepts such as the Larmor precession frequency, spin-spin, and spin-lattice relaxation times are discussed in some detail.

5.1 "CLASSICAL SPIN"

A direct experimental demonstration of the existence of an electron spin magnetic moment was first performed by Stern and Gerlach in 1921 by passing neutral silver atoms through an inhomogeneous magnetic field. In order to account for the fine structure of lines in the spectral series of some of the elements and also to account for the anomalous Zeeman effect, Uhlenbeck and Goudsmit introduced in 1925 the hypothesis that the electron rotates, or spins, about an axis like a top. Why an electron should have a spin as well as a charge is something that physics, as it is presently constituted at least, cannot answer. In any event, numerous experimental results can be explained by assuming the existence of an electron spin with an associated spin angular momentum $s\hbar$. Since a spinning charged body constitutes a current, the spinning motion should give rise to a magnetic field or a magnetic moment. According to Dirac's theory of the electron, an assumption of $s\hbar$ for the electron spin angular momentum leads to a spin magnetic moment μ_s given by

$$\mu_s = \left(\frac{-e}{m}\right)s\hbar \tag{5.1}$$

As we have already mentioned in connection with Eqs. (4.28) and (4.29), the magnitude of **s** is $\sqrt{\frac{1}{2}(\frac{1}{2} + 1)}$, or $\sqrt{3}/2$, but the projection of **s** along any given axis (defined by the direction of the static field, say) is only $\frac{1}{2}$. For this reason, the electron is said to possess a spin of $\frac{1}{2}$. Note also that the minus sign in Eq. (5.1) arises because the electron charge is negative; in this context, e is a positive quantity so that μ_s and **s** are oppositely directed.

Although the electron spin is clearly a quantum-mechanical attribute, its dynamical behavior can, in some respects, be described in terms of classical mechanics. For this reason and because most of us feel more comfortable with classical than with quantum mechanics, we shall first examine the dynamics of a "classical spin." If classical mechanics were applicable, when an electron spin is placed in a magnetic field H, it will be acted upon by a torque $\tau = \mu_0 \mu_s \times \mathbf{H}$. Now, τ is equal to the rate of change of angular momentum, which in this case is equal to $d(s\hbar)/dt$. Thus, the equation of motion for the electron spin is

$$\frac{d(s\hbar)}{dt} = \mu_0 \mu_s \times \mathbf{H} \tag{5.2}$$

Combining this equation with Eq. (5.1) we find the equation of motion for the spin magnetic moment:

$$\frac{d\mu_s}{dt} = \gamma_s \mu_s \times \mathbf{H} \tag{5.3}$$

where

$$\gamma_s = -\frac{\mu_0 e}{m} \tag{5.4}$$

When these equations are contrasted with those of Eqs. (4.13) and (4.14) for electron orbital motion, we notice that γ_s has twice the value of γ_c.

Let us now proceed to see what happens when a static field $\mathbf{H}_0$ is applied in the z direction. In this case Eq. (5.3) becomes

$$\frac{d\mu_s}{dt} = \gamma_s \mu_s \times \hat{\mathbf{z}} H_0 \tag{5.5}$$

If $\mu_s = (\hat{\mathbf{x}}\mu_{sx} + \hat{\mathbf{y}}\mu_{sy} + \hat{\mathbf{z}}\mu_{sz})e^{j\omega_L t}$, we obtain the following simultaneous equations for μ_{sx} and μ_{sy}:

$$j\omega_L \mu_{sx} = \gamma_s \mu_{sy} H_0 \tag{5.6}$$

$$j\omega_L \mu_{sy} = -\gamma_s \mu_{sx} H_0 \tag{5.7}$$

Solution of these equations yields the following relationships:

$$\mu_{sy} = \pm j\mu_{sx} \tag{5.8}$$

and

$$\omega_L = \pm \gamma_s H_0 \tag{5.9}$$

If initially μ_s makes an angle θ_0 with the z axis and its projection on the x-y plane makes an angle ϕ_0 with the x axis as shown in Fig. 5.1, then the angular dependence and time dependence of μ_s are given by

$$\mu_s = \mu_s \sin \theta_0 (\hat{\mathbf{x}} + \hat{\mathbf{y}} e^{-j\pi/2})e^{j(\omega_L t + \phi_0)} + \hat{\mathbf{z}}\mu_s \cos \theta \tag{5.10}$$

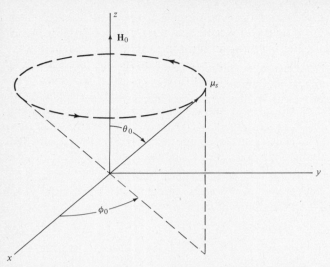

Figure 5.1   Larmor precession of the spin magnetic moment μ_s.

with $\omega_L = -\gamma_s H_0$ a positive quantity. In other words, the constant projection of μ_s on the x-y plane rotates clockwise with respect to the field direction at a constant frequency ω_L known as the Larmor frequency. Correspondingly, μ_{sz} is constant and has the value $\mu_s \cos \theta$.

Now let us see what happens when an alternating magnetic field **h** as well as a static field $\mathbf{H_0}$ is applied to the isolated spin, considered as a classical vector. Since ordinarily $H_0 \gg |\mathbf{h}|$, only components of $|\mathbf{h}|$ perpendicular to $\mathbf{H_0}$ will have any significant effect on the dynamics of μ_s. Therefore, with no loss of generality, we can consider **h** to lie in a plane perpendicular to $\mathbf{H_0}$. Furthermore, inasmuch as μ_s precesses about $\mathbf{H_0}$ with a constant frequency ω_L, it is expedient to decompose **h**, if linearly polarized, into two counterrotating components, one rotating clockwise and the other counterclockwise with respect to $\mathbf{H_0}$. If ω is nearly equal to the resonance of frequency ω_L, the component of **h** rotating in the direction opposite to that of μ_s will have relatively little effect on the dynamics of μ_s. This is because μ_s and the component of **h** under discussion will pass each other at a relative frequency of about 2ω, rendering any cumulative interaction highly unlikely. On the other hand, for the component of **h** rotating in the same direction as μ_s and with about the same frequency, cumulative interaction leading to a significant effect on the dynamics of the spin is quite plausible. This can best be demonstrated by transforming the equation of motion [Eq. (5.3)] written with respect to the laboratory frame to another frame of reference that rotates synchronously with the relevant component of **h** discussed above.

Before we embark on the mathematics of transformation, it would be instructive to surmise what type of results such a calculation is likely to yield. To begin with, both the static and time-varying fields should be time invariant in the rotating frame. The static field $\mathbf{H_0}$ remains static in the rotating frame since the frame rotates about the $\mathbf{H_0}$ axis, while the time-varying field should appear stationary to an observer rotating with the frame at the same frequency as the field **h** itself. It then follows that the magnitude and direction of the effective field $\mathbf{H}_{\mathrm{eff}}$, also static in the rotating frame, should be a function of H_0, ω, and $|\mathbf{h}|$. If ω and $|\mathbf{h}|$ are held constant, as is usually the case in a typical experiment, $\mathbf{H}_{\mathrm{eff}}$ should be a function of H_0. As H_0 is increased from a value below to a value above resonance, the direction of $\mathbf{H}_{\mathrm{eff}}$ as well as its magnitude can be expected to change significantly at or near resonance. Inasmuch as $\boldsymbol{\mu}_s$ precesses about $\mathbf{H}_{\mathrm{eff}}$, we should expect the axis of precession of $\boldsymbol{\mu}_s$ to also change with H_0. Indeed, if $\mathbf{H}_{\mathrm{eff}}$ should lie in a plane perpendicular to $\mathbf{H}_0$ for some appropriate value of H_0, $\boldsymbol{\mu}_s$, in its precessional cycle, could be expected alternately to align itself parallel and then antiparallel to $\mathbf{H}_0$.

Consider now a vector function of time $\mathbf{F}(t)$ with cartesian components $F_x(t)$, $F_y(t)$, and $F_z(t)$ along a set of rectangular coordinates, or

$$\mathbf{F} = \hat{\mathbf{x}}F_x + \hat{\mathbf{y}}F_y + \hat{\mathbf{z}}F_z \tag{5.11}$$

If the orientation of x, y, z as well as of F_x, F_y, F_z is a function of time, then the total time derivative of $\mathbf{F}$ is given by

$$\frac{d\mathbf{F}}{dt} = \hat{\mathbf{x}}\frac{dF_x}{dt} + F_x\frac{d\hat{\mathbf{x}}}{dt} + \hat{\mathbf{y}}\frac{dF_y}{dt} + F_y\frac{d\hat{\mathbf{y}}}{dt} + \hat{\mathbf{z}}\frac{dF_z}{dt} + F_z\frac{d\hat{\mathbf{z}}}{dt} \tag{5.12}$$

Now, let us specifically assume that the coordinate system x, y, z rotates with an angular velocity $\boldsymbol{\Omega}$ about some arbitrary axis. Then

$$\frac{d\hat{\mathbf{x}}}{dt} = \boldsymbol{\Omega} \times \hat{\mathbf{x}} \tag{5.13}$$

with similar expressions for $\hat{\mathbf{y}}$ and $\hat{\mathbf{z}}$. Substituting Eq. (5.13) into Eq. (5.12) and with the help of Eq. (5.11), we find

$$\frac{d\mathbf{F}}{dt} = \frac{\delta\mathbf{F}}{\delta t} + \boldsymbol{\Omega} \times \mathbf{F} \tag{5.14}$$

where the partial time derivative of $\mathbf{F}$, $\delta\mathbf{F}/\delta t$, is equal to $\hat{\mathbf{x}}\,\delta F_x/\delta t + \hat{\mathbf{y}}\,\delta F_y/\delta t + \hat{\mathbf{z}}\,\delta F_z/\delta t$.

Identifying $\mathbf{F}$ with $\boldsymbol{\mu}_s$ we can rewrite Eq. (5.3) in terms of a coordinate system rotating with an angular velocity $\boldsymbol{\Omega}$:

$$\frac{\delta\boldsymbol{\mu}_s}{\delta t} = \boldsymbol{\mu}_s \times (\gamma_s\mathbf{H} + \boldsymbol{\Omega}) \tag{5.15}$$

Thus we see that the motion of $\boldsymbol{\mu}_s$ in the rotating coordinate system is governed by an equation of the same form as in the laboratory system except that the actual field $\mathbf{H}$ must be replaced by an effective field

$$\mathbf{H}_{\mathrm{eff}} = \mathbf{H} - \frac{\boldsymbol{\Omega}}{|\gamma_s|} \tag{5.16}$$

where we have replaced γ_s by $-|\gamma_s|$ to emphasize that γ_s is negative. To appreciate the utility of Eq. (5.16), let us try to use it to find the motion of $\boldsymbol{\mu}_s$ in a static field $\mathbf{H} = \hat{\mathbf{z}}H_0$. If we take $\mathbf{\Omega}$ to be equal to $\hat{\mathbf{z}}|\gamma_s|H_0$, $\mathbf{H}_{eff}$ must be zero. It thus follows from Eq. (5.15) that $\delta\boldsymbol{\mu}_s/\delta t = 0$, indicating that $\boldsymbol{\mu}_s$ is stationary with respect to the rotating frame or, in other words, fixed with respect to $\hat{\mathbf{x}}$, $\hat{\mathbf{y}}$, $\hat{\mathbf{z}}$. The motion of $\boldsymbol{\mu}_s$ with respect to the laboratory frame is therefore that of a vector field in a set of axes that rotate at a frequency $\mathbf{\Omega}$. Equivalently, $\boldsymbol{\mu}_s$ rotates at an angular velocity $\mathbf{\Omega} = \hat{\mathbf{z}}|\gamma_s|H_0$ with respect to the laboratory. Note that Ω is equal to the Larmor frequency ω_L, which is obtained by a direct solution of Eq. (5.5), as it should be.

If $\mathbf{H}$, in addition to its static component $\hat{\mathbf{z}}H_0$, has also a time-varying component $\mathbf{h}(t)$, Eq. (5.3) becomes

$$\frac{d\boldsymbol{\mu}_s}{dt} = \gamma_s\boldsymbol{\mu}_s \times [\hat{\mathbf{z}}H_0 + \mathbf{h}(t)] \tag{5.17}$$

According to Eqs. (5.15) and (5.16), to obtain the effective field in a frame rotating with frequency ω about $\mathbf{H}_0$, H_0 in Eq. (5.17) should be replaced by $H_0 - \omega/|\gamma_s|$. Similarly, $\mathbf{h}(t)$ should be replaced by a static field $\hat{\mathbf{y}}h^1$; so long as the frame rotates at a frequency $\hat{\mathbf{z}}\omega$, the rotating field $\mathbf{h}(t)$ would appear to be stationary to an observer attached to the rotating frame. Accordingly, Eqs. (5.16) and (5.17) for this case take the specific forms

$$\frac{\delta\boldsymbol{\mu}_s}{\delta t} = \boldsymbol{\mu}_s \times \mathbf{H}_{eff} \tag{5.18}$$

where

$$\mathbf{H}_{eff} = \hat{\mathbf{z}}\left(H_0 - \frac{\omega}{|\gamma_s|}\right) + \hat{\mathbf{y}}h \tag{5.19}$$

Physically, these equations state that in the rotating frame the moment behaves as if it experienced an effective field $\mathbf{H}_{eff}$. $\boldsymbol{\mu}_s$ therefore precesses in a cone of fixed angle β_0 about the direction of $\mathbf{H}_{eff}$ and at an angular frequency $|\gamma_s|H_{eff}$. As we anticipated, H_{eff} is a function of H_0, ω, and h.

The dynamics of $\boldsymbol{\mu}_s$ at resonance is illustrated in Fig. 5.2 assuming that it is in the y-z plane and oriented at an angle θ_0 from the z direction at $t = 0$. We note that the motion of $\boldsymbol{\mu}_s$ is periodic; if it is initially oriented along the θ_0 direction, it periodically returns to that direction. If the system is exactly at resonance, i.e., $\omega = |\gamma_s|H_0$, $\mathbf{H}_{eff}$ is simply equal to $\hat{\mathbf{y}}h$. The locus of the tip of $\boldsymbol{\mu}_s$ will thus lie in a plane parallel to but at a distance $\mu_s \sin\theta_0$ from the x-z plane. Note also that below resonance $\mathbf{H}_{eff}$ has a negative $\hat{\mathbf{z}}$ component, whereas above resonance $\mathbf{H}_{eff}$ has a positive z component. These observations indicate that below resonance $\mathbf{H}_{eff}$ lies in the fourth quadrant of the y-z plane whereas above resonance it lies in the first quadrant of the y-z plane. In other words, the angle between $-\hat{\mathbf{z}}$ and $\mathbf{H}_{eff}$ increases with increasing H_0 (at fixes ω).

The above picture describes the situation in the reference frame rotating at a frequency ω about the $\mathbf{H}_0$ direction. It follows that in the laboratory frame

---

[1]The choice of $\hat{\mathbf{y}}$ rather than $\hat{\mathbf{x}}$, say, is entirely arbitrary and in no way affects the general results to follow.

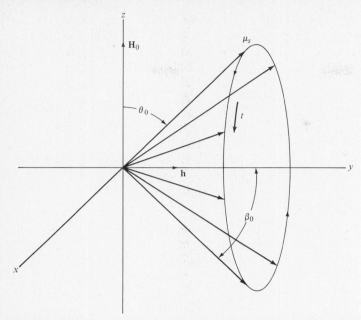

Figure 5.2 Locus of μ_s at resonance as time increases in a frame rotating at a frequency ω about the z axis.

the assembly of vectors, composed of μ_s and $\mathbf{H}_{\text{eff}}$, must in addition rotate at a frequency ω about the $\mathbf{H}_0$ or z direction. This situation, as seen by an observer in the laboratory, is roughly sketched in Fig. 5.3(a); the details of the locus of the tip of μ_s are dependent on the initial orientation of μ_s and the ratio of $|\gamma|H_{\text{eff}}$ and ω.

According to Eq. (5.19), the angle between $\mathbf{H}_{\text{eff}}$ and $\mathbf{H}_0$ decreases with increasing H_0 for given ω and h. Similarly, if H_0 and h are held fixed, this angle will increase with increasing ω. Although H_0 is usually the variable in a typical laboratory resonance experiment, it is nevertheless customary to assume that H_0 and h are fixed and ω is the variable.[2] With an initial value of ω far below that required for resonance and with no alternating field $\mathbf{h}$ applied, $\mathbf{H}_{\text{eff}}$, according to Eq. (5.19) lies along the z direction. If an isolated spin happens to be in the z direction when $\mathbf{H}_0$ is applied, μ_s will stay aligned parallel to $\hat{\mathbf{z}}$ as this is already the direction of minimum Zeeman energy $-\mu_s \cdot \mathbf{H}_0$. Then, if a rotating field $\mathbf{h}$ is applied at $t = 0$, $\mathbf{H}_{\text{eff}}$ will tip slightly away from the $\hat{\mathbf{z}}$ direction. We can thus expect μ_s to trace out a cone about the direction of $\mathbf{H}_{\text{eff}}$ with its half-cone angle corresponding to the angle between $\mathbf{H}_{\text{eff}}$ and $\mathbf{H}_0$. However, if $h \ll H_0$, this half angle will be very small, and for all practical purposes we can consider μ_s as being aligned with the new $\mathbf{H}_{\text{eff}}$ direction. Furthermore, it seems

---

[2]It is extremely difficult to construct oscillators with a constant output power over a broad band of frequencies. For this reason it is common practice to vary the magnitude of $\mathbf{H}_0$ for a fixed ω (and therefore a fixed oscillator output) from a value above to a value below resonance.

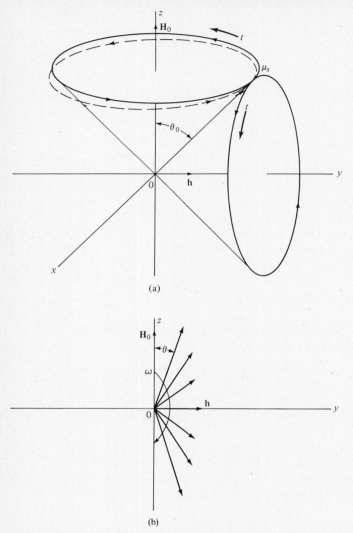

Figure 5.3 **(a)** Locus of μ_s at resonance as time increases in the stationary laboratory frame. The resultant locus of μ_s due to precession about $\mathbf{H}_{eff} = \mathbf{h}$ and $\mathbf{H}_0$ is shown dotted. **(b)** Locus of $\mathbf{H}_{eff}$ in the laboratory frame as ω increases.

reasonable to surmise that if ω is increased sufficiently slowly through resonance so that the direction of $\mathbf{H}_{eff}$ also changes slowly, μ_s will align with $\mathbf{H}_{eff}$ as we pass through resonance. As we increase ω from zero to infinity, we therefore should expect the angle θ between $\mathbf{H}_{eff}$ (and therefore μ_s) and $\mathbf{H}_0$ to increase from 0° to 180°, as depicted in Figure 5.3(b). In other words, whereas μ_s is oriented in the direction of $\mathbf{H}_0$ far below resonance, it is oriented perpendicular to $\mathbf{H}_0$ at resonance (where $\omega = |\gamma_s|H_0$) and antiparallel to $\mathbf{H}_0$ far above resonance. Indeed, if $h \ll H_0$ as is usually the case, resonance according to Eq. (5.19) should occur within a very small range of ω. In other words, in the limit of vanishingly small but finite h, $\boldsymbol{\mu}_s$ is aligned parallel to $\mathbf{H}_0$ below resonance, perpendicular to $\mathbf{H}_0$ exactly at resonance, and antiparallel to $\mathbf{H}_0$ above resonance: The spin (or $\boldsymbol{\mu}_s$) flips in direction at resonance.

Before we conclude our discussion of the dynamics of an isolated classical spin, it is illuminating to examine the problem briefly from an energy standpoint. First, since the spin is isolated from its environment, it cannot on the average absorb any energy from an alternating source. On the other hand, it is possible for it to receive energy from the source during one part of the cycle provided this energy is returned to the source during another part of the cycle. Referring to Fig. 5.2, we see that for a given $\mathbf{H}_{\text{eff}}$ the angle θ between $\boldsymbol{\mu}_s$ and $\mathbf{H}_0$ starting initially from θ_0 increases with time t until $t = \pi/|\gamma_s|h$, beyond which θ decreases back toward θ_0. Since the interaction energy between the moment and the fields is equal to $-\boldsymbol{\mu}_s \cdot \hat{\mathbf{z}}H_0$, the energy first increases and then decreases as $\boldsymbol{\mu}_s$ proceeds on its precessional cycle.[3] Thus, all the energy it takes to tilt $\boldsymbol{\mu}_s$ away from $\hat{\mathbf{z}}H_0$ is returned in a complete cycle of $\boldsymbol{\mu}_s$ around the cone. In other words, there is no absorption of energy from the alternating field, but energy is alternately received from and returned to it.

5.2 THE ISOLATED ELECTRON SPIN

In Section 5.1 we examined the behavior of a classical spin in some detail. Although it is well known that the spin of an electron obeys quantum-mechanical rather than classical laws, the dynamics of a classical spin as enumerated in Section 5.1 is nevertheless very similar in many respects to that of an electron spin governed by the laws of quantum mechanics. Inasmuch as classical mechanics is more akin to common sense, it is helpful to retain a physical picture of the behavior of a classical spin in magnetic fields *provided* we are mindful of the differences in behavior between the two cases. In what follows we shall emphasize the similarities and differences in the dynamics of these two cases.

To begin with, whereas the orientation of a classical spin can be arbitrary, the possible orientations of an actual electron spin are fixed by the rules of quantization in quantum mechanics. As we have mentioned previously, the projection of $\mathbf{s}$ along a given axis can only be $\pm\frac{1}{2}$.[4] Although the direction of this axis is in general arbitrary, its direction in practice is fixed by the nature of the experiment. For example, if the component of $\mathbf{s}$ along the direction of a measuring field $\mathbf{H}_0$ is to be determined, $\mathbf{H}_0$ defines the relevant direction under discussion.

Since, according to wave mechanics, the magnitude of $\mathbf{s}$ is $\sqrt{|s|(|s| + 1)}$ where $s = \pm\frac{1}{2}$ is the projection along a given axis, the angle that $\mathbf{s}$ makes with

---

[3]In the laboratory frame the interaction between $\boldsymbol{\mu}_s$ and $\hat{\mathbf{z}}H_0$ will cause $\boldsymbol{\mu}_s$ to have an additional precession about $\mathbf{H}_0$, as shown in Fig. 5.3(a). However, since the angle between $\boldsymbol{\mu}_s$ and $\mathbf{H}_0$ is constant for this superposed precession, the interaction energy $-\boldsymbol{\mu}_s \cdot \mathbf{H}_0$ is time independent.

[4]Actually, in quantum mechanics the spin is not a vector but an operator. Correspondingly, we speak of the expectation value of a spin along a given axis rather than the component of vector $\mathbf{s}$ along this axis. If $\mathbf{s}$ is treated as a vector, the treatment is known as semiclassical. For our purposes the semiclassical approach is quite sufficient.

$\mathbf{H}_0$ is $\theta_s = \cos^{-1}(s/\sqrt{|s|(|s| + 1)})$, which is either 54.7° or 123.3°. As $\boldsymbol{\mu}_s$ is oppositely directed to $\mathbf{s}$, it must make corresponding angles of 123.3° and 54.7° with $\mathbf{H}_0$. In a semiclassical sense, these quantized angles presumably correspond to those attained after equilibrium is established subsequent to the application of $\mathbf{H}_0$. Starting from an arbitrary initial angle with respect to $\mathbf{H}_0$, $\boldsymbol{\mu}_s$ will relax to the proper orientation after some elapse time. If the spin is truly isolated as assumed here, it will take an infinitely long time for this to occur. On the other hand, if the spin is coupled to its environment, this characteristic time will be finite. If the spin under consideration is part of an assembly of spins dispersed in a crystal lattice, it can be coupled to the environment by two distinct mechanisms: spin-spin or spin-lattice interactions. Corresponding to these two mechanisms, there are two relaxation times identified, respectively, as spin-spin and spin-lattice relaxation times τ_2 and τ_1. Note that for the situation under consideration, as $\boldsymbol{\mu}_s$ relaxes toward the proper angle with respect to $\mathbf{H}_0$, the interaction energy $-\mu_0 \boldsymbol{\mu}_s \cdot \mathbf{H}_0$ between $\boldsymbol{\mu}_s$ and $\mathbf{H}_0$ must change with time. In other words, for the realignment of $\boldsymbol{\mu}_s$ toward the quantized angles to occur, the spin assembly must transfer energy to and ultimately attain thermal equilibrium with the lattice. If this equilibrium is achieved directly, i.e., through spin-lattice interaction, the associated characteristic time is the spin-lattice relaxation time τ_1. If the equilibrium is established through spin-spin and then spin-lattice interaction, the spin-spin relaxation time τ_2 is involved. Although τ_1 and τ_2 take on very different values in different materials, they are not likely to be longer than a millisecond.

Now consider what happens if a rotating field $\mathbf{h}$ of frequency ω is applied after the equilibrium state in the presence of $\mathbf{H}_0$ described above has been established. First, with the application of $\mathbf{H}_0$, the original energy level E_0 of the electron is split into two levels, one at a value $\mu_B H_0$ above and one at a value $\mu_B H_0$ below E_0 where μ_B, the Bohr magneton, is the magnetic moment component of the electron spin along the $\mathbf{H}_0$ axis. As shown in Fig. 5.4, the z component of the spin $\mathbf{s}$ in the lower energy state is oriented antiparallel to $\mathbf{H}_0$ whereas in the upper energy state it is oriented parallel to $\mathbf{H}_0$. Correspondingly, the z component of the magnetic moment $\boldsymbol{\mu}_s$ is oriented upward and down-

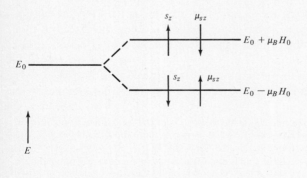

Figure 5.4 Energy level diagram of an electron spin in a static magnetic field.

ward in the lower and upper energy states, respectively. For simplicity we shall refer to the spin in the lower and upper energy states as *down* and *up*, respectively, although the spin **s** itself is not oriented along the z axis. In order for a spin to flip from the down to the up orientation, Fig. 5.4 indicates that an energy equal to $2\mu_B H_0$ must be supplied. This energy can be supplied by the rotating field **h** provided its frequency ω is related to this energy difference by the quantum-mechanical relation

$$\hbar\omega = 2\mu_B H_0 \tag{5.20}$$

corresponding to the condition of electron spin resonance. If $\omega \neq 2\mu_B H_0/\hbar$, no spin flip is possible as there are no permissible energy levels for the spin to flip to. We thus conclude that as ω is increased from zero through resonance and beyond, the down-spin orientation remains downward until resonance is reached at which time it flips upward. Likewise, the up-spin flips downward at resonance. As ω is increased beyond its value corresponding to resonance, no further spin-flipping activity occurs. This spin-flipping phenomenon at resonance is reminiscent of similar behavior for the classical spin discussed in Section 1. In the latter case abrupt flipping at resonance occurs provided $h \ll H_0$. Presumably, this is also true in the quantum-mechanical case; if the inequality $h \ll H_0$ does not hold, the interaction between μ_s and **h** will modify the energy level shown in Fig. 5.4. It should also be noted, however, that the upper and lower energy levels are infinitely sharp as implicitly assumed only if the spin-flipping time is infinite.[5] This is similar to requiring that the rate of change of H_0 through resonance be sufficiently small in the classical case.

In conclusion, we have found that the dynamics of the classical and electron spins are, in many respects, remarkably similar. For vanishingly small values of the r.f. field h, they both flip over an infinitesimal range of the driving frequency ω for a constant static field H_0 in a resonance experiment *provided* that ω is varied through resonance at an extremely slow rate (in the classical case) and the lifetime of the state approaches infinity (in the quantum-mechanical case.) Although the concept of spin quantization along a given axis is absent in classical mechanics, this difference appears inconsequential as far as their resonance dynamics is concerned *provided* only the component of **s** along the H_0 axis is considered in either case. Implicit in this last statement is the assumption that damping forces due to the coupling of the spin to its environment must exist to enable the spin system to attain its equilibrium state.

5.3 ELECTRON SPIN ASSEMBLY

In Section 5.2 we found that in order for a spin to attain its equilibrium state it cannot be truly isolated. Indeed, it is generally coupled in some way to other spins and to the lattice in any given sample. In this section we shall examine the modes of coupling and define the relaxation times involved.

---

[5]This statement follows from the uncertainty principle, which states that $\Delta E\, \Delta t \sim \hbar$ so that the uncertainty in the location of an energy level ΔE depends on the "lifetime" of the state Δt. If $\Delta t \to \infty$, $\Delta E \to 0$, and the energy level under consideration is infinitely sharp.

First, we should briefly examine the case where an atom has a spin of more than $\frac{1}{2}$. Under the Russel-Saunders type of coupling prevalent in magnetic materials, the various electron spins s_i in the atom will combine to form a resultant spin $\mathbf{S} = \sum_{i=1}^{n} \mathbf{s}_i$. The quantization condition for each spin is still applicable; that is, the projection of each spin along a given axis must still be $\pm\frac{1}{2}$. If S is the projection of $\mathbf{S}$ upon a given axis, the magnitude of $\mathbf{S}$ is $\sqrt{S(S + 1)}$ and the number of possible orientations is $2S + 1$. Thus, if there are n spins per atom, then $\mathbf{S} = n(\frac{1}{2})$ and the number of orientations $2n(\frac{1}{2}) + 1 = n + 1$. For these orientations the projection of $\mathbf{S}$ along a given axis will take on the values $n(\frac{1}{2})$, $(n - 2)(\frac{1}{2}), \ldots, (n - 2)(-\frac{1}{2})$, $n(-\frac{1}{2})$. At resonance any of these spins can be reversed; for every spin reversal, the total magnetic moment along the field direction changes by two Bohr magnetons. Correspondingly, the total spin $\mathbf{S}$ or the total magnetic moment $\boldsymbol{\mu}_s$ change their orientation to another allowable value. For the case where only a small amount of paramagnetic material is dissolved into a diamagnetic host [as in the gadolinium salt $Gd(C_2H_5SO_4)_3 \cdot 9H_2O$], the paramagnetic atoms are sufficiently far apart that the interaction between individual atomic spins is quite weak. In this case the atomic spin composed of several coupled electron spins can be considered to be more or less isolated. Thus, the conditions for the quantization of $\mathbf{S}$ just enumerated are applicable to every paramagnetic atom. As far as relaxation processes are concerned, the spin-spin relaxation effects can be neglected since the atomic spins are quite weakly coupled to each other. However, spin-lattice relaxation cannot be ignored because the paramagnetic spins are still coupled to the diamagnetic lattice. For substances with higher paramagnetic concentration, even the spin-spin coupling cannot be ignored. Although the relaxation mechanisms can be expressed in terms of the quantum-mechanical transition probabilities, the procedure is so involved that a phenomenological approach is frequently more fruitful.

We shall now proceed to show how the phenomenological spin-spin and spin-lattice relaxation times τ_2 and τ_1 are introduced into the Bloch equation formulation in connection with paramagnetic resonance. To begin with, let us rewrite the equation of motion [Eq. (5.3)] for an atomic moment $\boldsymbol{\mu}_S$ rather than for the moment $\boldsymbol{\mu}_s$ of a single-electron spin. Thus,

$$\frac{d\boldsymbol{\mu}_S}{dt} = \gamma_S \boldsymbol{\mu}_S \times \mathbf{H} \tag{5.21}$$

where, according to Eqs. (4.29) and (5.4), $\gamma_S = \gamma_s = -\mu_0 e/m$. Multiplying both sides of Eq. (5.21) by N and noting that $N\boldsymbol{\mu}_s$ is equal to the magnetization $\mathbf{M}$ (magnetic moment per volume), we have

$$\frac{d\mathbf{M}}{dt} = \gamma_S \mathbf{M} \times \mathbf{H} \tag{5.22}$$

Since the torque acting on $\mathbf{M}$, namely, $\mathbf{M} \times \mathbf{H}$, is perpendicular to both $\mathbf{M}$ and $\mathbf{H}$, it cannot change the angle between $\mathbf{M}$ and $\mathbf{H}$. This implies that it is not possible to align $\mathbf{M}$ with $\mathbf{H}$ or to magnetize a paramagnetic material by the application of a field, an assertion clearly in contradiction with experiment. One way to resolve this dilemma is to add to the right-hand side of Eq. (5.22) a damping term that will, among other things, allow the alignment of **H** with **M** to occur in due course. Bloch proposed the following generalization of Eq. (5.22)[6]:

$$\frac{dM_\perp}{dt} = \gamma \, (\mathbf{M} \times \mathbf{H})_\perp - \frac{M_\perp}{\tau_2} \qquad (5.23)$$

$$\frac{dM_z}{dt} = \gamma \, (\mathbf{M} \times \mathbf{H})_z - \frac{M_z - M_0}{\tau_1} \qquad (5.24)$$

where $\gamma = -g\mu_0 e/2m$ and g is the Landé g-factor given by Eq. (4.31). The subscript z refers to the vector component parallel to z or $\mathbf{H}_0$, while the subscript $\perp$ refers to components perpendicular to $\mathbf{H}_0$. Note that two characteristic times, one for $M_z - M_0$ and another for the transverse component, have been introduced into the Bloch equations.

To appreciate the physical significance of the characteristic times τ_1 and τ_2, let us see what happens to $M_\perp$ and M_z as a function of time if the static and r.f. magnetic fields are removed. In that case, the $\mathbf{M} \times \mathbf{H}$ terms in Eq. (5.23) and (5.24) vanish and the solution of these equations is given by

$$M_z = M_0 - \Delta M e^{-t/\tau_1} \qquad (5.25)$$

$$M_\perp = M_{\perp 0} e^{-t/\tau_2} \qquad (5.26)$$

where $\Delta M = \Delta(M_0 - M_z)$ at $t = 0$ is assumed positive. We thus see that both M_z and $M_\perp$ change with time. Whereas M_z changes from an initial value $M_0 - \Delta M$ to M_0 with a time constant τ_1, $M_\perp$ decays as t approaches infinity from an initial value $M_{\perp 0}$ to zero with a time constant τ_2. Thus, we can expect the magnitude of **M** $(=\sqrt{M_z^2 + M_\perp^2})$ to change with time. Accordingly, the locus of the tip of **M** should form an inward spiral toward the z or $\mathbf{H}_0$ direction, as sketched in Fig. 5.5.

Recalling that the Zeeman energy density $-\mathbf{M} \cdot \mathbf{H}_0$ is dependent on the z component but not the $\perp$ component of **M**, we see from Eqs. (5.25) and (5.26) that the change in M_z changes the energy of the spin system but the decay of $M_\perp$ does not. The extra energy gain by M_z during decay is presumably absorbed by the environment, chiefly the lattice. In quantum-mechanical terms this extra energy must excite quanta of lattice vibrations or phonons if the spin system are to attain thermal equilibrium with the lattice. For this reason, the time constant τ_1 is called the spin-lattice relaxation time. On the other hand, τ_2, representing the characteristic relaxation time required for the $\perp$ component of the spins in the system to lose phase coherence between them, is known as the spin-spin relaxation time. In this connection, we note that the projections of all spins in the sample rotate in phase in response to the r.f. driving field; after this field is removed, the various spins will in due course precess at various phases with respect to each other.

[6]F. Bloch, *Phys. Rev.* **70**, 460 (1946).

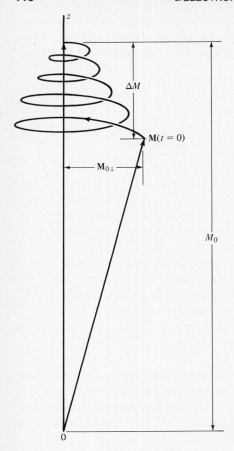

Figure 5.5 Locus of **M** as a function of time.

In quantum-mechanical terms τ_1 and τ_2 can be visualized as follows. τ_1 can be attributed to either a direct or an indirect process. In the direct process the spin, in the simple case of $s = \frac{1}{2}$, makes a transition from the state $m_s = \frac{1}{2}$ to the $m_s = -\frac{1}{2}$ state, thereby giving up an energy of $g\mu_B H_0$, according to Eq. (4.34); for the case of an isolated spin these energy levels correspond to those depicted in Fig. 5.4. Simultaneous with the spin flip, a lattice mode of frequency $\omega_k = g\mu_B H_0/\hbar$ makes a transition from the state n_k to the state $n_k + 1$; in particle language, we say that a phonon (quantum of lattice vibration) is created. In an indirect process two lattice modes rather than one participate in the transition process. Simultaneous with the spin flip, one phonon of wave number k is absorbed while a phonon with wave number k' is emitted. Energy conservation is satisfied by having $\hbar(\omega_{k'} - \omega_k) = g\mu_B H_0$. It may be instructive to give an example of the direct relaxation process. Consider the dipolar magnetic field acting on a given atom due to the presence of another atom residing somewhere in the lattice. Since the strength of this field depends on the distances between the atoms, it will be modulated by lattice vibrations. The part of the lattice vibration spectrum with Fourier components at $\omega \approx g\mu_B H_0/\hbar$ can cause transitions of the atom from the $m_s = \frac{1}{2}$ to the $m_s = -\frac{1}{2}$ state and vice versa. Thus, energy exchange between the system in the lattice is facilitated.

As stated earlier, the spin-spin relaxation time τ_2 characterizes the rate of decay of the transverse magnetization $M_\perp$. This decay is due to the loss of phase coherence between the transverse magnetic moment components of the constituent spins. In a homogeneous spin system this loss of coherence is due to a two-spin relaxation process in which one makes a transition from the state $m_s = \frac{1}{2}$ to the state $m_s = -\frac{1}{2}$ while another simultaneously makes a transition in the opposite direction, i.e., from the $m_s = -\frac{1}{2}$ state to the $m_s = \frac{1}{2}$ state. In other words, while one spin flips down another flips up. Inasmuch as the z component of $\mathbf{M}$ is not disturbed in this process, no energy is exchanged between the spin system and the lattice; the simultaneous spin flip merely interrupts transverse precessional coherence without contributing to the attainment of thermal equilibrium of the spin system. For this reason, the spin-spin relaxation time is characterized by a time constant τ_2 differing in general from the spin-lattice relaxation time τ_1.

It is not easy to understand in quantum-mechanical terms how phase coherence between spins is interrupted by a simultaneous flipping of spins. Clearly, to do so, we must resort to time-dependent quantum mechanics; the mathematical complexities involved are often so substantial that the real physics behind the calculation is obscure. For our purposes, it is probably best to obtain some insight in this regard by looking at the motion of a classical spin during reversal. In this connection, we recall that in Section 5.1 it was found that during its reversal a classical spin continuously changes its rate and axis of precession with time. This behavior is, in fact, well summarized by Eq. (5.19):

$$\mathbf{H}_{\text{eff}} = \hat{\mathbf{z}}\left(H_0 - \frac{\omega}{|\gamma|}\right) + \hat{\mathbf{y}}h \tag{5.19}$$

As ω is varied through its resonance value $|\gamma|H_0$, the direction of $\mathbf{H}_{\text{eff}}$ is seen to change from being parallel, to perpendicular, and finally to antiparallel to $\hat{\mathbf{z}}H_0$. At the same time, the magnitude of $\mathbf{H}_{\text{eff}}$ also changes as ω changes, implying the rate of precession also changes. This change in both the rate and the axis of spin precession during reversal clearly will cause the flipping spins to precess in phases different from the stationary ones, thus leading to phase incoherence.

5.4 EQUATION OF MOTION FOR MAGNETIZATION

Equations (5.23) and (5.24), the Bloch equations of motion for the magnetization developed for paramagnetic resonance, are not the only ones in use in microwave magnetics. Since the damping terms are phenomenological in character, there is a possibility that other similar but different damping terms can be introduced instead. Indeed, whereas the Bloch equations (5.23) and (5.24) are used extensively in paramagnetic resonance, the Landau-Lifshitz equation[7]

$$\frac{d\mathbf{M}}{dt} = \gamma(\mathbf{M} \times \mathbf{H}) - \frac{\lambda}{M^2} \mathbf{M} \times (\mathbf{M} \times \mathbf{H}) \tag{5.27}$$

is most often used in ferromagnetic resonance. In common with the damping terms of the Bloch equations, the damping term of the Landau-Lifshitz equation in the absence of r.f. fields will also cause $\mathbf{M}$ to align with H_0. The rate of decay in this case is related to the damping frequency λ.[8] On the other hand, whereas the magnitude of $\mathbf{M}$ is not conserved during the decay process in the Bloch formulation, it is definitely conserved in the Landau-Lifshitz formulation. The latter can be verified by noting that in Eq. (5.27) the vector $\mathbf{M} \times (\mathbf{M} \times \mathbf{H})$ is always perpendicular to $\mathbf{M}$ and thus can change the direction but not the magnitude of $\mathbf{M}$. Unlike the case of a paramagnet, the local magnetic moments in a ferromagnet are held together by extremely strong exchange forces. Thus, the assumption of the conservation of M during decay seems justified. However, experimental evidence in this regard is far from conclusive. For this and other reasons, both the Bloch and Landau-Lifshitz equations are used in ferromagnetic resonance. Note also that only one damping constant is used in the Landau-Lifshitz equation, while two relaxation times are used in the Bloch equations.

Another damping form has been proposed by Gilbert.[9] The Gilbert equation of motion for the magnetization is

$$\frac{d\mathbf{M}}{dt} = \gamma(\mathbf{M} \times \mathbf{H}) + \frac{\alpha}{M} \mathbf{M} \times \frac{d\mathbf{M}}{dt} \tag{5.28}$$

Although it can be shown by vector transformation that under certain circumstances Eqs. (5.27) and (5.28) are equivalent, the behavior of $\mathbf{M}$ obtained from the two equations may in general differ. In particular, we note that in the Gilbert case the damping of $\mathbf{M}$ is proportional to the rate of change of the magnetization of $d\mathbf{M}/dt$, while in the Landau-Lifshitz case it is not explicitly dependent upon this rate. In any event, it can be shown that the Bloch equations (5.23) and (5.24), the Landau-Lifshitz equation (5.27), and the Gilbert equation (5.28) are all equivalent for the case of small r.f. drive ($h \ll H_0$) provided τ_2 is set equal to $1/\omega\alpha$ and $\alpha^2 \ll 1$.

We have thus far elaborated on three equations of motion for the magnetization. At this point it may be philosophically satisfying to surmise what the most general damping form would be like. In this regard, it seems that three damping terms, one proportional to $\mathbf{M}$, another proportional to $\mathbf{M} \times \mathbf{H}$, and a third proportional to $\mathbf{M} \times (\mathbf{M} \times \mathbf{H})$, would be general enough. Thus, we can

[7]L. Landau and L. Lifshitz (in English), *Phys. Z. Sowjetunion* **8**, 153 (1935).

[8]λ, a positive quantity, can be related to a damping parameter α by the equation $\lambda = -\gamma\alpha M$. Thus, the second term in Eq. (5.27) can also be written $(\gamma\alpha/M)\mathbf{M} \times (\mathbf{M} \times \mathbf{H})$.

[9]T. A. Gilbert, "Equation of Motion of Magnetization," Armour Research Foundation Rep. No. 11, Chicago, II (January 25, 1955).

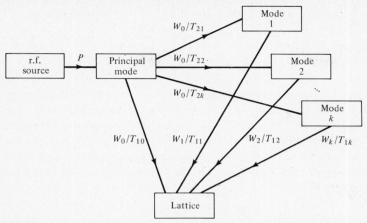

Figure 5.6 Schematic representation of the energy transfer process between the uniform precession mode, the spin modes, and the lattice. [After Fletcher, Le Graw, and Spencer, *Phys. Rev.* 117, 955 (1960).]

write

$$\frac{d\mathbf{M}}{dt} = \gamma(\mathbf{M} \times \mathbf{H}) + \alpha_1 \mathbf{M} + \alpha_2 \mathbf{M} \times \mathbf{H} + \alpha_3 \mathbf{M} \times (\mathbf{M} \times \mathbf{H}) \qquad (5.29)$$

For this equation to be useful, α_1, α_2, and α_3 must be determined experimentally or by relating them to quantum-mechanical transition probabilities. Thus, a general equation like Eq. (5.29) is only of academic interest.

Perhaps the most illuminating derivation of an equation of motion for the magnetization is that given by Fletcher, LeGraw, and Spencer.[10] Using the rate of energy transfer between the uniform precession, the spin waves, and the lattice, they developed a phenomenological description of electron-spin relaxation in ferromagnetic insulators. This procedure leads to an equation of motion for $\mathbf{M}$ containing the relaxation time of the uniform precession to the lattice τ_{10}, the relaxation time of the uniform precession to the kth spin mode τ_{2k}, and the relaxation time of the kth spin wave to the lattice τ_{1k}. Each spin wave state is treated separately; it is not necessarily at equilibrium with other spin wave states, or with the uniform precession, or with the lattice.

The relaxation process involved in the calculation is depicted in Fig. 5.6. In this figure W_0 represents the energy of the principal mode directly excited by the r.f. field while W_k represents the energy of the kth spin mode. P is defined as the net power per unit volume absorbed by the sample. From the calculations indicated above the following equations of motion were found:

$$\frac{d}{dt}(M_x^2 + M_y^2) = \frac{2M_0 P}{H_0 + (N_T - N_z)M_0} - (M_x^2 + M_y^2)\left(\frac{1}{\tau_{10}} + \sum_k \frac{1}{\tau_{2k}}\right) \qquad (5.30)$$

[10]R. C. Fletcher, R. C. LeGraw, and E. G. Spencer, *Phys. Rev.* **117**, 955 (1960).

and

$$\frac{d}{dt}(M_0 - M_z) = \frac{P}{H_0 + (N_T - N_z)M_0} - \frac{M_x^2 + M_y^2}{2M_0}\left(\frac{1}{\tau_{10}} - \frac{1}{\tau_{1k}}\right) - \frac{M_0 - M_z}{\tau_{1k}}$$

(5.31)

where M_x, M_y, and M_z are, respectively, the x, y, and z components of the magnetization. M_0 is the saturation magnetization, and $\mathbf{H}_0$ is the applied static magnetic field. N_T and N_z are, respectively, the demagnetizing factors transverse and parallel to the direction of $\mathbf{H}_0$.

To facilitate comparison with the phenomenological Bloch equations (5.23) and (5.24), let us rewrite Eqs. (5.30) and (5.31) in such a way as to isolate the damping terms involved:

$$\frac{dM_\perp}{dt} = \gamma(\mathbf{M} \times \mathbf{H})_\perp - \frac{M_\perp}{2}\left(\frac{1}{\tau_{10}} + \sum_k \frac{1}{\tau_{2k}}\right)$$

(5.32)

$$\frac{dM_z}{dt} = \gamma(\mathbf{M} \times \mathbf{H})_z + \frac{M_\perp^2}{2M_0}\left(\frac{1}{\tau_{10}} - \frac{1}{\tau_{1k}}\right) - \frac{M_z - M_0}{\tau_{1k}}$$

(5.33)

where

$$\gamma(\mathbf{M} \times \mathbf{H})_\perp = \frac{M_0 P}{M_\perp[H_0 + (N_T - N_z)M_0]}$$

(5.34)

and

$$\gamma(\mathbf{M} \times \mathbf{H})_z = -\frac{P}{H_0 + (N_T - N_z)M_0}$$

(5.35)

Equation (5.32) is seen to be the exact equivalent of the Bloch equation (5.23) for transverse relaxation provided

$$\frac{1}{\tau_2} = \frac{1}{2}\left(\frac{1}{\tau_{10}} + \sum_k \frac{1}{\tau_{2k}}\right)$$

(5.36)

Furthermore, if $\tau_{10} = \tau_{1k}$ for a given k, the second term on the right-hand side of Eq. (5.33) vanishes and this equation for longitudinal relaxation is also identical with that of Bloch. If $\tau_{10} \neq \tau_{1k}$, the Bloch equations can be shown to still give a description of the motion of the system that has most of the essential features provided we find the appropriate expression for τ_1.[11] Thus, it is very comforting to note that the time-honored Bloch equations can be deduced from basic energy considerations involving a transfer of energy between the uniform precession and spin modes, between uniform precession and lattice, and between spin modes and lattice. Of course, these equations are linearized; no reaction of the spin modes to the uniform mode has been assumed. Furthermore, the actual relaxation process is probably more complicated than that depicted in Fig. 5.6. In particular, the r.f. field could excite any one or a combination of the spin modes, and these may excite still others instead of relaxing directly to the lattice.

[11]Ibid.

As we have noted, the Bloch, Landau-Lifshitz, and Gilbert equations are all equivalent for the case of small damping and small r.f. signal levels. Under these conditions, we now see that for all practical purposes they are also equivalent to the Fletcher-LeGraw-Spencer equations.

PROBLEMS

5.1 For $f = 10^{10}$ GHz, $|\gamma_s| = 2.8$ MHz/Oe, $h = 10^{-3}$ Oe, find the precessional frequency of μ_s about H_{eff}.

5.2 **(a)** For small signal and small losses ($\alpha^2 \ll 1$), show that the Bloch-Bloembergen equation is the same as the Landau-Lifshitz and Gilbert equations provided $\tau_2 = 1/\omega\alpha$.

    (b) Show that the magnitude of **M** in the Landau-Lifshitz and Gilbert equations is conserved.

5.3 Find the general expression for the dotted resultant locus of μ_s shown in Fig. 5.3(a).

5.4 **(a)** Using Eqs. (5.25) and (5.26), find the magnitude of **M** as a function of time.

    (b) For $M_{\perp 0} = 0.1M_0$, $\Delta M = 0.005M_0$, $\tau_2 = 10^{-8}$ s and $\tau_1 = 10^{-7}$ s, plot M/M_0 as a function of time.

5.5 Suppose five electron spins in an atom are coupled together via the Russel-Saunders coupling. Find the minimum allowable angle between the resultant spin and any given axis.

Chapter 6

Paramagnetic Resonance and Applications

In Chapter 5 we examined the dynamics of electron spins in combined static and time-varying magnetic fields. In this chapter we shall apply the basic equations of motion for the magnetization developed there to the problem of resonance in paramagnetic materials. We shall show how the spin-spin and spin-lattice relaxation times τ_1 and τ_2 can be experimentally evaluated. The relationship between τ_1, τ_2 and the measured linewidth ΔH as well as the sources of contribution to ΔH and characteristic line shapes will also be considered. The application of paramagnetic resonance theory to masers and lasers will also be discussed.

6.1 PARAMAGNETIC RESONANCE

6.1.1 Equation of Motion

In Chapter 5 we stated the Bloch equations of motion for the magnetization:

$$\frac{dM_\perp}{dt} = \gamma(\mathbf{M} \times \mathbf{H})_\perp - \frac{M_\perp}{\tau_2} \qquad (5.23)$$

$$\frac{dM_z}{dt} = \gamma(\mathbf{M} \times \mathbf{H})_z - \frac{M_z - M_0}{\tau_1} \qquad (5.24)$$

where $M_\perp$ and M_z are, respectively, the components of $\mathbf{M}$ perpendicular and parallel to the applied static field. Correspondingly, $(\mathbf{M} \times \mathbf{H})_\perp$ and $(\mathbf{M} \times \mathbf{H})_z$ are, respectively, the torque components perpendicular and parallel to the static field. τ_2 is the transverse or spin-spin relaxation time while τ_1 is the longitudinal or spin-lattice relaxation time.

As the discussion in connection with the Bloch equations in Chapter 5 indicates, the relaxation terms introduced are plausible but not necessarily exact. For example, whereas it seems logical enough that the longitudinal magnetization M_z ought to relax toward its terminal equilibrium value M_0, it is not at all certain that in a material composed of an incredible number of atoms this relaxation can be characterized by a single relaxation time τ_1. The same can be said of τ_2. Thus, we should perhaps think of τ_1 and τ_2 as composite times rather than as times characteristic of a single process. Furthermore, one can further contend that perhaps not every relaxation process is describable in terms of a characteristic decay time. In any event simplicity of the Bloch damping forms can render the related equations of motion for the magnetization very useful in paramagnetic and nuclear magnetic resonance if τ_1 and τ_2 can be simply and unambiguously related to easily measurable experimental parameters.

In a paramagnetic resonance experiment τ_1 and τ_2 are not determined directly. Instead, the resonance linewidth ΔH defined as the field separation between the half-power points (referred to the power absorption at resonance) is actually measured. In what follows we shall show how ΔH is related to τ_1 and τ_2 under different experimental conditions.

6.1.2 Resonance in Weak r.f. Fields

Let us assume that **H** is composed of a static and time-varying component:

$$\mathbf{H} = \hat{\mathbf{z}}H_0 + \mathbf{H}_\perp(t) = \hat{\mathbf{z}}H_0 + h(\hat{\mathbf{x}} - \hat{\mathbf{y}}j)e^{j\omega t} \tag{6.1}$$

where the time-varying field is assumed to be circularly polarized and rotating clockwise about $\hat{\mathbf{z}}$, the direction of H_0; if $h \ll H_0$ as is implicitly assumed here, any time-varying component in the direction of H_0 will have negligible effect. For an electron spin a positive ω corresponds to the direction of Larmor precession and possible resonance. Conversely, a negative ω corresponds to the direction opposite to the Larmor precession and no resonance can be expected in this latter case. Corresponding to the form of **H** assumed in Eq. (6.1), there is a similar expression for **M**:

$$\mathbf{M} = \hat{\mathbf{z}}M_0 + \mathbf{M}_\perp(t) = \hat{\mathbf{z}}M_0 + m(\hat{\mathbf{x}} - \hat{\mathbf{y}}j)e^{j(\omega t + \theta)} \tag{6.2}$$

where θ is the angle between **m** and **h**. For the steady-state case considered here θ is expected to be independent of time. Substituting Eqs. (6.1) and (6.2) into Eqs. (5.23) and (5.24), we have

$$j\omega m e^{j\theta} = j\gamma M_0 h - j\gamma H_0 m e^{j\theta} - \frac{m}{\tau_2} e^{j\theta} \tag{6.3}$$

$$M_z \simeq M_0 \tag{6.4}$$

The second equation was obtained by neglecting second-order terms, i.e., terms involving m^2 as well as the products of m and h. This is justified for the cases where $h \ll H_0$ and $m \ll M_0$. Equating the real and imaginary components on both sides of Eq. (6.3), we finally obtain

$$\tan \theta = -\frac{1}{(\omega_0 - \omega)\tau_2} \tag{6.5}$$

$$m = \frac{|\gamma| M_0 \tau_2 h}{(\omega_0 - \omega)\tau_2 \cos \theta - \sin \theta} \tag{6.6}$$

where $\omega_0 = |\gamma| H_0$. Noting that $\sin \theta = \tan \theta / \sqrt{1 + \tan^2 \theta}$ and $\cos \theta = \sin \theta / \tan \theta$, we easily find that[1]

$$\sin \theta = -\frac{1}{[1 + (\omega_0 - \omega)^2 \tau_2^2]^{1/2}} \tag{6.7}$$

and

$$\cos \theta = \frac{(\omega_0 - \omega)\tau_2}{[1 + (\omega_0 - \omega)^2 \tau_2^2]^{1/2}} \tag{6.8}$$

Substituting Eqs. (6.7) and (6.8) into Eq. (6.6), we obtain the relationship between m and h:

$$m = \frac{|\gamma| M_0 \tau_2}{[1 + (\omega_0 - \omega)^2 \tau_2^2]^{1/2}} h \tag{6.9}$$

6.1.3 Power Absorption

We shall begin our discussion of power absorption by first deriving the expression for the power absorbed by a specimen in a magnetic field directly from Maxwell's equations. Taking the dot product of **H** and both sides of Eq. (2.1) and the dot product of **E** and both sides of Eq. (2.2), we have

$$\mathbf{H} \cdot \nabla \times \mathbf{E} = -\mathbf{H} \cdot \frac{\partial \mathbf{B}}{\partial t} \tag{6.10}$$

and

$$\mathbf{E} \cdot \nabla \times \mathbf{H} = \mathbf{E} \cdot \mathbf{i} + \mathbf{E} \cdot \frac{\partial \mathbf{D}}{\partial t} \tag{6.11}$$

Subtracting Eq. (6.11) from Eq. (6.10) and using the vector identity $\mathbf{A} \cdot (\nabla \times \mathbf{B}) - \mathbf{B} \cdot (\nabla \times \mathbf{A}) = \nabla \cdot (\mathbf{B} \times \mathbf{A})$ where **A** and **B** are any two vectors, we find

$$-\mathbf{H} \cdot \frac{\partial \mathbf{B}}{\partial t} - \mathbf{E} \cdot \frac{\partial \mathbf{D}}{\partial t} - \mathbf{E} \cdot \mathbf{i} = \nabla \cdot (\mathbf{E} \times \mathbf{H}) \tag{6.12}$$

Integrating over the volume V, Eq. (6.12) becomes

$$\int_V \left(\mathbf{H} \cdot \frac{\partial \mathbf{B}}{\partial t} + \mathbf{E} \cdot \frac{\partial \mathbf{D}}{\partial t} + \mathbf{E} \cdot \mathbf{i} \right) dv = -\oint_S (\mathbf{E} \times \mathbf{H}) \cdot ds \tag{6.13}$$

where we have also used Gauss's theorem [Eq. (2.44)]. For our purposes here, it is expedient to relate **B** to **H** and **M** and correspondingly **D** to **E** and **P** by

---

[1] We have associated the minus sign with $\sin \theta$ rather than with $\cos \theta$ so that the power absorption given by Eq. (6.17) is positive.

means of the constitutive relations (2.10) and (2.11)

$$\mathbf{B} = \mu_0(\mathbf{H} + \mathbf{M}) \tag{6.14}$$

and
$$\mathbf{D} = \epsilon_0\mathbf{E} + \mathbf{P} \tag{6.15}$$

where $\mathbf{P}$ is the polarization. Substituting Eqs. (6.14) and (6.15) into Eq. (6.13), we find

$$\int_v \left[\frac{\partial}{\partial t}\left(\frac{\mu_0 H^2 + \epsilon_0 E^2}{2} \right) + \mathbf{E} \cdot \mathbf{i} + \mu_0 \mathbf{H} \cdot \frac{\partial \mathbf{M}}{\partial t} + \mathbf{E} \cdot \frac{\partial \mathbf{P}}{\partial t} \right] dv = -\oint_s (\mathbf{E} \times \mathbf{H}) \cdot d\mathbf{s} \tag{6.16}$$

The first term on the left-hand side of Eq. (6.16) represents the rate of change of stored energy density in the magnetic and electric fields,[2] while the second term is the familiar ohmic term representing energy loss per unit volume in the form of heat per unit time. The third term, the one of most interest to us, represents the energy per unit volume absorbed per unit time by the magnetization $\mathbf{M}$ placed in a magnetic field $\mathbf{H}$. The fourth term represents the energy per unit volume absorbed per unit time by the polarization $\mathbf{P}$ placed in an electric field $\mathbf{E}$. The net increase in stored energy density, heat loss, and power absorbed by $\mathbf{M}$ and $\mathbf{E}$ just enumerated must be supplied externally. Thus, the right-hand side of Eq. (6.16) must represent the energy flow *into* the volume V across the bounding surface S per unit time; $\mathbf{E} \times \mathbf{H}$ is just the familiar Poynting vector.

From the above discussion we find that the power absorbed by a magnetic sample per unit volume is given by

$$P = \mu_0 \mathbf{H}_\perp \cdot \frac{d\mathbf{M}}{dt} \tag{6.17}$$

From Eqs. (6.1) and (6.2), P is found to be given by[3]

$$P = -\omega\mu_0 hm \sin \theta \tag{6.18}$$

Substituting the expressions for $\sin \theta$ and m from Eqs. (6.7) and (6.9), we have

$$P = \frac{\mu_0 \omega |\gamma| M_0 \tau_2}{1 + (\omega_0 - \omega)^2 \tau_2^2} h^2 \tag{6.19}$$

In the steady state the power P absorbed by the sample per unit volume appears as lattice vibrations.

It is easier to vary H_0 than ω in the laboratory since the former can be varied by merely changing the current through the electromagnet supplying the field. On the other hand, for cavity-type microwave oscillators such as the reflex klystron, the electronic tuning range is quite small. For microwave oscillators with a periodic structure such as the backward wave oscillator (BWO), the oscillation frequency can be changed by merely changing the beam voltage.[4] However, the power output from the BWO is not constant with frequency; thus for

---

[2]R. F. Soohoo, *Microwave Electronics*, Addison-Wesley, Reading, MA, 1971, p. 50.

[3]As usual, since Eq. (6.17) involves the product of $\mathbf{H}$ and $\mathbf{M}$, we must use here the real part of $\mathbf{H}$ and $\mathbf{M}$ given by Eqs. (6.1) and (6.2) rather than their complex counterparts.

[4]R. F. Soohoo, *Microwave Electronics*, p. 121.

the results to be meaningful, some form of automatic level control (ALC) must be used. Furthermore, if ω is to be varied, cavities that enhance the sample absorption by Q ($10^3 - 10^4$ at microwave frequencies) cannot be used conveniently. For these reasons the linewidth is usually measured in terms of field rather than frequency.

A linewidth ΔH defined as the field separation between the half-power points on the P-H_0 plot can now be defined. Equating the expression for P at some values of H_0 [as given by Eq. (6.19)] to one-half its corresponding value at resonance; i.e., at $\omega = \omega_0$, we have

$$\frac{\mu_0 \omega |\gamma| M_0 \tau_2 h^2}{1 + (\omega_{02,1} - \omega)^2 \tau_2^2} = \tfrac{1}{2} \mu_0 \omega |\gamma| M_0 \tau_2 h^2 \tag{6.20}$$

Solving for ω_0, we find

$$\omega_{02,1} = \omega \pm \frac{1}{\tau_2} \tag{6.21}$$

at which the power absorption is down to one-half its value at resonance. Here the plus sign corresponds to H_{02} while the minus sign corresponds to H_{01}. The full linewidth ΔH is then given by

$$\Delta H = \frac{2}{|\gamma| \tau_2} \tag{6.22}$$

6.1.4 Magnetic Susceptibility

$\mathbf{M}_\perp$ can be related to $\mathbf{H}_\perp$ by the following relation:

$$\mathbf{M}_\perp = \chi \mathbf{H}_\perp \tag{6.23}$$

where χ is the susceptibility. As can be seen from Eq. (6.5), the angle θ between $\mathbf{M}_\perp$ and $\mathbf{H}_\perp$ is in general nonzero. Consequently, although $\mathbf{M}_\perp$ and $\mathbf{H}_\perp$ both lie in the x-y plane, they are not necessarily in time phase. This implies that the susceptibility χ in Eq. (6.23) must in general be complex. Substituting the expressions for $\mathbf{H}_\perp$ and $\mathbf{M}_\perp$ given by Eqs. (6.1) and (6.2) into Eq. (6.23) we have

$$\chi = \frac{m}{h} (\cos \theta + j \sin \theta) \tag{6.24}$$

Substituting Eqs. (6.7)–(6.9) for $\sin \theta$, $\cos \theta$, and m into Eq. (6.24) and letting $\chi = \chi' - j\chi''$, we obtain the following expression[5] for the real and imaginary components of χ:

$$\chi' = \frac{|\gamma| M_0 (\omega_0 - \omega) \tau_2^2}{1 + (\omega_0 - \omega)^2 \tau_2^2} \tag{6.25}$$

$$\chi'' = \frac{|\gamma| M_0 \tau_2}{1 + (\omega_0 - \omega)^2 \tau_2^2} \tag{6.26}$$

---

[5]Note that χ is set equal to $\chi' - j\chi''$ rather than $\chi' + j\chi''$ to render χ'' given by Eq. (6.26) always positive.

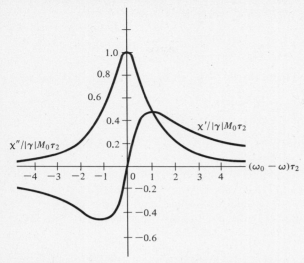

Figure 6.1 Normalized susceptibility components vs normalized frequency deviation from resonance.

These expressions were derived for $\mathbf{H}_\perp$ and $\mathbf{M}_\perp$ rotating about the positive z direction. $\mathbf{H}_\perp$ and $\mathbf{M}_\perp$ are referred to as positive circularly polarized. For negative circularly polarized $\mathbf{H}_\perp$ and $\mathbf{M}_\perp$, we need only change the sign of ω in Eqs. (6.25) and (6.26). $\chi'/|\gamma|M_0\tau_2$ and $\chi''/|\gamma|M_0\tau_2$ are plotted against $(\omega_0 - \omega)\tau_2$ in Fig. 6.1 for the positive circularly polarized case, i.e., with ω positive. The same quantities can be plotted for the negative circularly polarized case, with ω negative.[6] Notice that, whereas resonance is exhibited in the first case, as manifested by the peaking of χ'' at $\omega_0 = \omega$, none is evident in the second case.

If, in a given experiment, both senses of polarization of $\mathbf{H}_\perp$ are present, the total power absorption by the sample will be dependent on the relative values of the r.f. magnetic fields associated with the two senses of polarization. That this is the case can be easily demonstrated by a comparison of Eq. (6.19) for power absorption and (6.26) for χ'', yielding the relation

$$P = \mu_0 \omega \chi'' h^2 \tag{6.27}$$

Whereas Eqs. (6.19) and (6.26) were derived for the positive circularly polarized magnetic field h_+, they are also applicable to the negative circularly polarized field h_-. It thus follows that the total power absorption by a sample in an elliptically polarized field is given by

$$P = \mu_0 \omega (\chi''_+ h_+^2 + \chi''_- h_-^2) \tag{6.28}$$

[6]This follows from the fact that if ω is changed in sign, $\mathbf{H}_\perp(t)$ and $\mathbf{M}_\perp(t)$ in Eqs. (6.1) and (6.2) will represent negative instead of positive circularly polarized fields.

Here χ''_+ is given by Eq. (6.26), while χ'' is given by

$$\chi''_- = \frac{|\gamma|M_0\tau_2}{1 + (\omega_0 + \omega)^2\tau_2^2} \tag{6.29}$$

Similarly, χ'_- can be obtained from Eq. (6.25) by changing the sign of ω, yielding

$$\chi'_- = \frac{|\gamma|M_0(\omega_0 + \omega)\tau_2^2}{1 + (\omega_0 + \omega)^2\tau_2^2} \tag{6.30}$$

In the present context ω is positive in the four equations (6.25), (6.26), (6.29), and (6.30).

Figure 6.1 shows universal curves for χ' and χ'' in that they are applicable for any values of ω, ω_0, and τ_2. In practice, ω is in the microwave band, so that ω and ω_0 can be expected to take on values in the order of $2\pi \times 10^{10}$ Hz, say. For typical paramagnetic materials, τ_2 may range from 10^{-10} to 10^{-9} s.[7]

6.2 HIGH-POWER SATURATION

The discussion in Section 6.1 concerned paramagnetic resonance in low signal levels. If the r.f. signal is sufficiently high, nonlinear terms in the Bloch equations of motion (5.23) and (5.24) can no longer be neglected. In particular if the inequality $h \ll H_0$ does not hold, we can expect $\mathbf{M}$ to deviate significantly in direction from that of $\mathbf{H}_0$. In this case M_z will have a time-varying component as well as a static component. Accordingly, in contrast to the low-power case, the spin-lattice relaxation time τ_1 as well as the spin-spin relaxation time τ_2 should also enter into the results of our calculation.

The best way to solve the Bloch equations (5.23) and (5.24) when h is not necessarily much less than H_0 is to transform these equations into a frame rotating synchronously with the r.f. circularly polarized field. In this rotating frame the effective field $\mathbf{H}_{\text{eff}}$ is given by Eq. (5.19):

$$\mathbf{H}_{\text{eff}} = \hat{\mathbf{z}}\left(H_0 + \frac{\omega}{\gamma}\right) + \hat{\mathbf{x}}h \tag{5.19}$$

Correspondingly, the Bloch equations in the rotating frame take on the following form:

$$\frac{dM_x}{dt} = \gamma(\mathbf{M} \times \mathbf{H}_{\text{eff}})_x - \frac{M_x}{\tau_2} \tag{6.31}$$

$$\frac{dM_y}{dt} = \gamma(\mathbf{M} \times \mathbf{H}_{\text{eff}})_y - \frac{M_y}{\tau_2} \tag{6.32}$$

$$\frac{dM_z}{dt} = \gamma(\mathbf{M} \times \mathbf{H}_{\text{eff}})_z - \frac{M_z - M_0}{\tau_1} \tag{6.33}$$

[7]A. H. Morrish, *The Physical Principles of Magnetism*, Wiley, 1965, p. 105.

where M, M_x, M_y, M_z represent magnetization and magnetization components in the rotating frame. Substituting the expression for $\mathbf{H}_{eff}$ given by Eq. (5.19) into these equations, we finally obtain

$$\frac{dM_x}{dt} = \gamma \left(H_0 + \frac{\omega}{\gamma} \right) M_y - \frac{M_x}{\tau_2} \tag{6.34}$$

$$\frac{dM_y}{dt} = \gamma \left[-\left(H_0 + \frac{\omega}{\gamma} \right) M_x + hM_z \right] - \frac{M_y}{\tau_2} \tag{6.35}$$

$$\frac{dM_z}{dt} = -\gamma hM_y - \frac{M_z - M_0}{\tau_1} \tag{6.36}$$

The general solution of Eqs. (6.34)–(6.36) is a sum of decreasing exponential terms and a steady-state solution obtained by setting dM_x/dt, dM_y/dt, and dM_z/dt equal to zero. The steady-state solution is given by

$$M_x = \frac{\gamma^2 (H_0 + \omega/\gamma) \tau_2^2 h}{1 + \gamma^2 (H_0 + \omega/\gamma)^2 \tau_2^2 + \gamma^2 \tau_1 \tau_2 h^2} M_0 \tag{6.37}$$

$$M_y = \frac{\gamma \tau_2 h}{1 + \gamma^2 (H_0 + \omega/\gamma)^2 \tau_2^2 + \gamma^2 \tau_1 \tau_2 h^2} M_0 \tag{6.38}$$

$$M_z = \frac{1 + \gamma^2 (H_0 + \omega/\gamma)^2 \tau_2^2}{1 + \gamma^2 (H_0 + \omega/\gamma)^2 \tau_2^2 + \gamma^2 \tau_1 \tau_2 h^2} M_0 \tag{6.39}$$

Equations (6.37)–(6.39) give magnetization components M_x, M_y, M_z in the rotating frame. To obtain the corresponding components $M_{x'}$, $M_{y'}$, $M_{z'}$ in the laboratory frame, we refer to the relationship between the coordinate systems shown in Fig. 6.2(a). From this figure we immediately arrive at the following

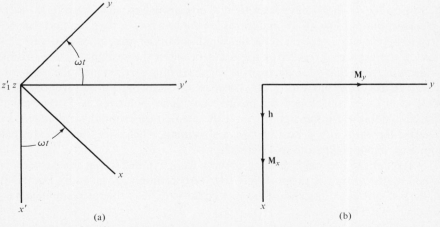

(a)   (b)

Figure 6.2 (a) Relationship between the rotating frame (x, y, z) and the laboratory frame (x', y', z'). (b) Relationship of $\mathbf{M}_x$, $\mathbf{M}_y$, and $\mathbf{h}$ in the rotating frame.

equations of transformation:

$$M_{x'} = M_x \cos \omega t - M_y \sin \omega t \tag{6.40}$$

$$M_{y'} = M_x \sin \omega t + M_y \cos \omega t \tag{6.41}$$

$$M_{z'} = M_z \tag{6.42}$$

Combining Eqs. (6.40)–(6.42) with Eqs. (6.37)–(6.39), we finally obtain the magnetization components in the laboratory frame of reference:

$$M_{x'} = \frac{\gamma(H_0 + \omega/\gamma)\tau_2 \cos \omega t - \sin \omega t}{1 + \gamma^2(H_0 + \omega/\gamma)^2\tau_2^2 + \gamma^2\tau_1\tau_2 h^2} \, \gamma h\tau_2 M_0 \tag{6.43}$$

$$M_{y'} = \frac{\gamma(H_0 + \omega/\gamma)\,\tau_2 \sin \omega t + \cos \omega t}{1 + \gamma^2(H_0 + \omega/\gamma)^2\tau_2^2 + \gamma^2\tau_1\tau_2 h^2} \, \gamma h\tau_2 M_0 \tag{6.44}$$

$$M_{z'} = \frac{1 + \gamma^2(H_0 + \omega/\gamma)^2\tau_2^2}{1 + \gamma^2(H_0 + \omega/\gamma)^2\tau_2^2 + \gamma^2\tau_1\tau_2 h^2} \, M_0 \tag{6.45}$$

The susceptibility in this high-power case can be calculated by a consideration of the phase relationship between $\mathbf{M}$ and $\hat{\mathbf{x}}h$. With the help of Eq. (6.24) and Fig. 6.2(b), we find the expression for the complex susceptibility χ:

$$\chi = \frac{M}{h} = \frac{M_x}{h} + j\frac{M_y}{h} \tag{6.46}$$

Setting χ equal to $\chi' - j\chi''$, we have

$$\chi' = \frac{\gamma^2(H_0 + \omega/\gamma)\tau_2^2}{1 + \gamma^2(H_0 + \omega/\gamma)^2\tau_2^2 + \gamma^2\tau_1\tau_2 h^2} \, M_0 \tag{6.47}$$

$$\chi'' = \frac{-\gamma\tau_2}{1 + \gamma^2(H_0 + \omega/\gamma)^2\tau_2^2 + \gamma^2\tau_1\tau_2 h^2} \, M_0 \tag{6.48}$$

where Eqs. (6.37)–(6.39) have been used. Comparing Eqs. (6.47) and (6.48) with the corresponding Eqs. (6.25) and (6.26) for the low-power case, we find that an extra term, $\gamma^2\tau_1\tau_2 h^2$, appears in the denominator in the present case. As we shall show, the presence of this extra term will broaden the resonance line. In this connection recall that γ for electrons is negative; thus, a resonance phenomenon is evident in Eqs. (6.47) and (6.48). These expressions and those given by Eqs. (6.43)–(6.45) were developed for positive circularly polarized $\mathbf{h}$. For negative circularly polarized $\mathbf{h}$ the sign for ω in these equations should be changed. In this case the ratio ω/γ is positive and no resonance behavior can be expected from Eqs. (6.34)–(6.45) and (6.47)–(6.48).

To find the linewidth ΔH, we observe from Eq. (6.27) that the power absorbed by the sample is proportional to χ''. Accordingly, χ'' given by Eq. (6.48) at the half-power points should be just equal to $\frac{1}{2}$, its value at resonance, i.e., at $\omega = |\gamma|H_0$. Thus,

$$\frac{-\gamma\tau_2 M_0}{1 + \gamma^2(H_{01,2} + \omega/\gamma)^2\tau_2^2 + \gamma^2\tau_1\tau_2 h^2} = \frac{1}{2}\left(\frac{-\gamma\tau_2 M_0}{1 + \gamma^2\tau_1\tau_2 h^2}\right) \tag{6.49}$$

where H_{02} and H_{01} are the field values corresponding to half-power points, one located above and one below resonance. Solving for $H_{01,2}$ in Eq. (6.49), we find

$$H_{01,2} = \frac{\omega}{|\gamma|} \pm \frac{\sqrt{1 + \gamma^2\tau_1\tau_2 h^2}}{|\gamma|\tau_2} \tag{6.50}$$

where the plus and minus signs are associated with H_{02} and H_{01}, respectively. It follows that the linewidth $\Delta H = H_{02} - H_{01}$ is given by

$$\Delta H = \frac{2}{|\gamma|\tau_2} \sqrt{1 + \gamma^2\tau_1\tau_2 h^2} \tag{6.51}$$

For small values of h such that $\gamma^2\tau_1\tau_2 \ll 1$, ΔH reduces to $2/|\gamma|\tau_2$, consistent with the value obtained for the low-power case given by Eq. (6.22). However, at high power levels, ΔH is broadened by the factor $\sqrt{1 + \gamma^2\tau_1\tau_2 h^2}$. Inasmuch as τ_2 can be determined by measuring the linewidth at low signal levels, τ_1 can also be found by measuring ΔH at high signal levels. The phenomenon associated with the broadening of ΔH at high signal levels is known as saturation.

Whereas τ_1 represents the characteristic time required for the spin system to attain thermal equilibrium with the lattice, τ_2 is associated with the destruction of phase coherence within the spin system itself. In contrast to the former case, no energy exchange occurs between the spin system and its environment in the latter case.

6.3 LINEWIDTH CONTRIBUTIONS

In Section 6.2 we showed that the linewidth ΔH is in general related to the relaxation times τ_1 and τ_2. Here we shall discuss the common line shapes that can occur in paramagnetic resonance as well as the possible contributions to the width of such lines.

6.3.1 Spin-Lattice Contributions

The linewidth of a given paramagnetic material is determined in part by the nature and strength of interaction between an electron spin and its surroundings. To begin with, a paramagnetic ion in a dilute paramagnet is subjected to a ligand field due to the matrix of diamagnetic ions in which it is embedded in a crystal lattice. The simplest method of treating the ligand field is to consider it as a purely electrostatic interaction. The ligands are regarded as charged ions located on given lattice points, which give rise to a crystalline potential that reflects the local symmetry of the environment resulting in energy level splittings for the magnetic electrons. The populations of these levels can be altered by transitions between levels. Such transitions can only be induced by thermal fluctuations, i.e., by the molecular motion within the matrix. In the case of a solid this motion consists of lattice vibrations, which may or may not be localized, together with, in a conducting solid, the kinematic effects associated with movement of the conduction electrons and, more rarely, of charged ions. The characteristic time associated with the foregoing phenomenon is just the so-called spin-lattice relaxation time τ_1.

For a simple two-level spin system the appropriate rate equation for the processes producing thermal equilibrium can be derived by a consideration of electron spin in a static magnetic field. Associated with the electron spin **s** there is a magnetic moment $\boldsymbol{\mu}$ proportional to **s** and directed opposite to it. Since the energy of interaction between the magnetic moment $\boldsymbol{\mu}$ and the static field $\mathbf{H}_0$ is $-\boldsymbol{\mu} \cdot \mathbf{H}_0$ and the electron spin could have one of two possible projections along the static field axis, i.e., $\pm \frac{1}{2}$, the original energy level of the electron E_0 is split into two levels separated by $2|\mu_z|H_0$ where μ_z is the component of $\boldsymbol{\mu}$ in the direction of $\mathbf{H}_0$.

Consider a number of uncoupled electron spins such as those existing in a dilute paramagnetic sample and let the populations in the $-$ spin and $+$ spin states be N_1 and N_2, respectively. Also let P_{12} be the probability per unit time that an electron in state 1 makes a transition to state 2 because of the interaction of the electron with its surroundings; conversely, let P_{21} be the probability per unit time that an electron in state 2 makes a transition to state 1 because of the interaction of the electron with its surroundings. At thermal equilibrium we should expect the rate of upward transition to be exactly equal to that of downward transition:

$$N_1^e P_{12} = N_2^e P_{21} \tag{6.52}$$

Equation (6.52) exhibits the *principle of detailed balance*, which in reality is merely a mathematical statement of thermal equilibrium in which the number of upward transitions is exactly canceled by the number of downward transitions so that the population of both states is independent of time.[8] It now follows from the Boltzmann distribution law, which states that $N_i^e \propto e^{-E_i/kT}$, and Eq. (6.52) that

$$\frac{P_{12}}{P_{21}} = e^{-(E_2 - E_1)/kT} \tag{6.53}$$

In order that the population of the system may reach the equilibrium distribution given by the Boltzmann distribution law from some nonequilibrium initial distribution or vice versa, there must exist a mechanism for inducing transitions between N_1 and N_2, which arises because of the coupling of the spins to some other system. It would be instructive to calculate the final population and the characteristic time required to reach the final state in terms of quantities P_{12}, P_{21}, N_1, and N_2 already introduced.

To begin with, let us replace N_1 and N_2 by two new variables n and N:

$$\begin{aligned} N &= N_1 + N_2 \\ n &= N_1 - N_2 \end{aligned} \tag{6.54}$$

Solving for N_1 and N_2, we find

$$N_1 = \tfrac{1}{2}(N + n)$$
$$N_2 = \tfrac{1}{2}(N - n) \tag{6.55}$$

Note that N_1 and N_2 are not in general equal to N_1^e and N_2^e, the values at thermal equilibrium. Indeed, the rate equations[9] are

$$\frac{dN_1}{dt} = N_2 P_{21} - N_1 P_{12}$$
$$\frac{dN_2}{dt} = N_1 P_{12} - N_2 P_{21} \tag{6.56}$$

Using Eq. (6.55) and noting that $N = N_1 + N_2 = N_1^e + N_2^e$ is independent of time, one sees that Eq. (6.56) becomes

$$\frac{dn}{dt} = N(P_{21} - P_{12}) - n(P_{21} + P_{12}) \tag{6.57}$$

which can be rewritten

$$\frac{dn}{dt} = \frac{n_0 - n}{\tau_1} \tag{6.58}$$

where

$$n_0 = N\left(\frac{P_{21} - P_{12}}{P_{21} + P_{12}}\right)$$
$$\frac{1}{\tau_1} = P_{21} + P_{12} \tag{6.59}$$

The solution of Eq. (6.58) is

$$n = n_0 + Ae^{-t/\tau_1} \tag{6.60}$$

where A is an arbitrary constant. Thus, n_0 represents the thermal equilibrium population difference while τ_1 is a characteristic time associated with the approach to thermal equilibrium, or the *spin-lattice relaxation time*. For example, if a sample is initially unmagnetized, the magnetization process can be described by the expression

$$n = n_0(1 - e^{-t/\tau_1}) \tag{6.61}$$

since $n = 0$ at $t = 0$. From Eq. (6.61) we see that τ_1 is a time constant characterizing the time required to magnetize an initially demagnetized sample.

We can readily find the relationship between Eq. (6.58) and its classical counterpart, i.e., the relaxation term for M_z in the Bloch equation (5.24). Multiplying both sides of Eq. (6.58) by the magnetic moment μ, we have

$$\frac{d(\mu n)}{dt} = \frac{\mu n_0 - \mu n}{\tau_1} \tag{6.62}$$

---

[9] We have implicitly assumed that the values of P_{12} and P_{21} to be used here are the same as those corresponding to thermal equilibrium [see Eq. (6.52)]. This is permissible providing the system at all times does not deviate much from thermal equilibrium.

Since $n = N_1 - N_2$ represents the population difference between levels 1 and 2 at a given time t while n_0 represents the thermal equilibrium population difference, we conclude that $\mu n_0 = M_0$ and $\mu n = M_z$. Thus, Eq. (6.62) can be rewritten to read

$$\frac{dM_z}{dt} = -\frac{M_z - M_0}{\tau_1} \tag{6.63}$$

It is comforting to note that this equation is exactly the same as the phenomencological Bloch equation, (5.24), for M_z with no transverse driving field $\mathbf{H}_\perp$.

If an alternating magnetic field $\mathbf{H}_\perp$ is applied to the sample in addition to the static field, further transitions between the $+\frac{1}{2}$ and $-\frac{1}{2}$ spin states will occur. Let C_{12} denote the probability per unit time of inducing the transition of an electron from state 1 to state 2, and let C_{21} denote that of the reverse process. Then, the complete rate equations become

$$\frac{dN_1}{dt} = N_2 P_{21} - N_1 P_{12} + N_2 C_{21} - N_1 C_{12}$$

$$\frac{dN_2}{dt} = N_1 P_{12} - N_2 P_{21} + N_1 C_{12} - N_2 C_{21} \tag{6.64}$$

Using the transformations of Eq. (6.55), Eq. (6.64) becomes

$$\frac{dn}{dt} = -2Cn + \frac{n_0 - n}{\tau_1} \tag{6.65}$$

where we have set $C_{12} = C_{21} = C$. The equality of C_{12} and C_{21} can be shown on very general grounds.[10]

At steady state the populations of states 1 and 2 remain unchanged with time. It follows that in this case $dn/dt = d(N_1 - N_2)/dt = 0$. Thus, we find the expression for n at statistical steady state from Eq. (6.65) as

$$n = \frac{n_0}{1 + 2C\tau_1} \tag{6.66}$$

The induced transition probability C, as might be expected, is linearly proportional to the r.f. energy density or power. Thus, we see from Eq. (6.66) that so long as $H_\perp$ is sufficiently small so that $2C\tau_1 \ll 1$, $n \simeq n_0$—or the population of the states is essentially equal to that at thermal equilibrium.

The rate of absorption of energy, dE/dt, is given by

$$\frac{dE}{dt} = (N_1 - N_2)C\hbar\omega = \hbar Cn\omega \tag{6.67}$$

where $N_1 C$ is the number of electron spins per unit time that go from the lower energy state to the higher one, $N_2 C$ is the number of electron spins per unit time that go in the reverse direction, and $\hbar\omega$ is the energy of a photon

[10]R. F. Soohoo, *Microwave Electronics*, p. 263.

corresponding to a frequency ω. It follows that $N_1 Ch\omega$ corresponds to energy absorption whereas $N_2 Ch\omega$ corresponds to energy emission. Thus, a positive dE/dt represents net energy absorption, whereas a negative dE/dt represents net energy emission. According to Eq. (6.66), $n > 0$ because n_0, C, and τ_1 are all positive. Consequently not only is $N_2^e < N_1^e$ at thermal equilibrium according to the Boltzmann distribution law, but $N_2 < N_1$, even at statistical steady state in the presence of an alternating magnetic field.

Before we proceed further, we might combine Eqs. (6.66) and (6.67) to obtain a final expression for energy absorption dE/dt:

$$\frac{dE}{dt} = n_0 Ch\omega \, \frac{1}{1 + 2C\tau_1} \tag{6.68}$$

So long as $2C\tau_1 \ll 1$, we see that dE/dt is proportional to C or to the r.f. power. However, if the r.f. power is sufficiently high that C is comparable to $\frac{1}{2}\tau_1$, the absorption levels off despite further increase in the r.f. power. This effect is known as saturation, a phenomenon that was discussed in a classical context in Section 6.2.

Next we shall delineate the possible contribution to τ_1 (and therefore to ΔH) and examine their relative importance. Several known mechanisms contribute to τ_1:[11]

1. For *direct interaction* between the spin system and the thermal electromagnetic radiation bath, the energy density in the bath at the frequencies and temperature of interest is much too small to produce relaxation times of the right order of magnitude.

2. For *indirect interaction* between the spin system and the "phonon radiation bath," the energy in the bath is approximately $(c/v)^3$ times that of the "photon radiation bath" of contribution 1. Since the velocity of sound v in a solid is of the order of 3×10^3 m/s, the phonon energy density is greater by a factor of about 10^{15}. This can easily outweigh the fact that the interaction between spins and phonons involves mechanisms that are indirect and inherently weaker than the direct magnetic resonance interaction between spins and photons. The first discussion of such mechanisms, by Waller, was based on modulation of the spin-spin interaction by the phonons, which induces oscillatory components in the distance between paramagnetic ions.[12] Two Waller processes can be distinguished—the direct and the Raman processes. In the direct processes a phohon of the same energy $\hbar\omega$ as the spin quantum required for a resonance transition is absorbed by the spin system, resulting in an up transition or a phonon emitted accompanied by a down transition. In the Raman process a phonon

[11]A. Abragan and P. Bleaney, *Electron Paramagnetic Resonance of Transition Ions*, Clarendon, Oxford, 1970, p. 60.

[12]I. Waller, *Z. Phys.* **79**, 370 (1932).

of any frequency ω_p may interact with a spin, causing an up or down transition within the spin system, the phonon being scattered with different frequencies $\omega_p - \omega$ and $\omega_p + \omega$, respectively. Although the relaxation time of Waller processes is considerably smaller than that of contribution 1, it is still appreciably longer than that experimentally observed.

3. A more potent mechanism and one which, unlike that of Waller, is independent of the degree of concentration of the magnetic ions is *modulation of the ligand field* by the lattice vibrations. This produces primarily a fluctuating electric field, which modulates the orbital motion of the magnetic electrons. Modulation of the orbital motion is transmitted to the spin by spin-orbit coupling. The foundations of the theory underlying these mechanisms were laid by Kronig[13] and Van Vleck[14] and have been extended by many others, notably Orbach.[15] As in contribution 2, these processes can be divided into the direct and Raman ones. In addition, the Orbach process involves absorption of a phonon by a direct process to excite the spin system to a much higher level, followed by the emission of another phonon of slightly different energy so that the magnetic ion is indirectly transferred from one level to the other of the ground doublet.

In the discussion of the rate at which the spin system relaxes to the phonons, we implicitly assumed that phonons are always in thermal equilibrium with both. If this were not the case, the energy that the phonons received from the spin system would quickly heat the resonant phonons to a temperature close to that of the spin system. Then the combined system (spin + phonon) would relax for a bath with time constant much longer than that for the phonon alone. This situation is likely to occur in the direct process, in which the number of participating phonons is small compared to the number of spins; only these phonons, which are at the same frequency as the resonance frequency (or roughly within the resonance linewidth), can interact with the spins. For this reason the heat capacity of the spin system is much higher than that of the phonons.

6.3.2 Spin-Spin Relaxation Time

So far we have examined the interaction between the spin system and the lattice but not the interaction between the paramagnetic ions. The latter, known generally as the spin-spin interaction, can arise in a number of ways, as will now be delineated.[16]

[13]R. De L. Kronig, *Physicas Grav.* **6**, 33 (1939).

[14]J. H. Van Vleck, *Phys. Rev.* **57**, 426 (1940).

[15]R. Orbach, *Proc. R. Soc. London Ser. A* **264**, 458, 485 (1961).

[16]Abragan and Bleaney, *Electron Paramagnetic Resonance*, p. 52.

Magnetic Dipole-Dipole Interaction. This arises from the influence of the magnetic field of one paramagnetic ion on the dipole moments of neighboring paramagnetic ions. This type of interaction is the simplest type of spin-spin interaction and the only one readily amenable to calculation. The actual local field at a given site will clearly depend on the location and orientation of the neighboring dipoles. Since in the paramagnetic state the dipoles are rather randomly oriented because of the influence of thermal agitation, there is a randomness to the local field in both magnitude and direction. This randomness in the local field gives rise to line broadening in a paramagnet.

The magnitude of this random field can be estimated by means of Eq. (4.63). For example, according to this equation, the magnitude of the magnetic field **H** at a perpendicular distance r from a dipole μ is equal to $\mu/4\pi r^3$. If μ is equal to that of an electron spin or a Bohr magneton and r is 6 or 7 Å, typical of the distance between neighboring paramagnetic ions in a normally paramagnetic salt, H will be on the order of 50 Oe. If each ion has a number of paramagnetic neighbors of higher magnetic moment, the net local field at a given site may be as high as 100–1000 Oe. If an external field H_0 is applied, the total local field at a given site will be equal to the vector sum of H_0 and **H**. Since H is usually small compared to the H_0 required for paramagnetic resonance at microwave frequencies (several kiloersteds), only the component of **H** parallel to H_0 will be significant. Since the size of this component is expected to change from site to site, it should give rise to a total field directed along H_0 that varies from site to site. This in turn will result in a random displacement of the resonance frequency of each ion, similar in effect to the case of a paramagnetic sample in an inhomogeneous applied field. For this reason this phenomenon is known as *inhomogeneous broadening*. Whereas the resonance frequency of each ion is displaced, the lifetime of the ion in a given quantum state is not altered. Therefore, there is no direct relationship between line broadening and relaxation time in this case. In other words, although every process with a finite relaxation time should give rise to line broadening, not all line broadening can be attributed to relaxation processes.

Relaxation Time Due to Spin Flip. If the paramagnetic ions are identical, so that they precess at the same frequency in the applied magnetic field, there is an additional resonance interaction. In this case the precessing components of one magnetic dipole set up an oscillatory field at another dipole that is just at the right frequency to cause magnetic resonance transitions, and vice versa. Consider, for example, the case of two spins i and j in a two-level system. If spins i and j are in the lower and upper energy states, respectively, then spin i can be thought of as pointing "down" while spin j is pointing "up." Because of the resonance interaction under discussion, spin i goes up into the upper state while spin j drops down into the lower state. Correspondingly, spin i flips up while spin j flips down. Clearly, then, this spin flip shortens the lifetime of the individual ion in a given quantum state and broadens the resonance line. This broadening may be on the order of 50% in magnitude of the inhomogeneously broadened linewidth. Note also that whereas the flipping of spin i increases the energy of the system by an amount $2\mu_B H_0$, the flipping of spin j decreases it by $2\mu_B H_0$. Thus, the total energy of the spin system remains unchanged, a phenomenon characteristic of spin-spin relaxation. Correspondingly, a relaxation time τ_2' can be related to the line broadening $(\Delta H)'$. From Eq. (6.22) we have

$$\hbar(\Delta\omega)' = g\mu_B(\Delta H)' \tag{6.69}$$

and τ_2' can be related to $(\Delta\omega)'$ by

$$\tau_2' = \frac{2}{(\Delta\omega)'} \tag{6.70}$$

Whatever process the line shape may be attributed to, it is of course possible to loosely define a time τ_2' by the relation $\tau_2' = 2/(\Delta\omega)'$. If τ_2' is attributable to a *real* relaxation process, then the relationship between τ_2' and the experimentally determined $(\Delta\omega)'$ or $(\Delta H)'$ has a definite physical basis. This process of line broadening is known as *homogeneous broadening*. On the other hand, if τ_2' is not a true relaxation time as in the case of a homogeneous broadening, then the relationship between τ_2' and $(\Delta\omega)'$ is still expedient but has little physical significance.

Interactions Associated with Electric Moments. In ions that have orbital momentum, the distribution of electric charge in the magnetic substates will be anisotropic. The charge distribution may possess an electric quadrupole or higher moments and, in some rare cases, an electric dipole moment. Such moments will interact with the electrostatic potential component associated with similar moments on adjacent ions, but the magnitude of the interaction is difficult to estimate. In some cases the electrostatic interaction may outweigh the magnetic interaction.

Exchange Interaction. In many paramagnetic compounds in which the separation of the magnetic ions is less than about 5 Å, exchange interaction between neighbors exceeds the purely dipole interaction. Exchange interaction can be further defined into two classes: isotropic exchange and anisotropic exchange. In the former case the line is narrowed in the center and extended in the wings if the spins are identical. The half-power width is reduced, and for this reason, the phenomenon is referred to as *exchange narrowing*. The narrowing process resembles that found in nuclear magnetic resonance in liquids: Through exchange interaction with other neighbors, any given neighboring dipole changes its orientation at a rate of the order of J/h [derivable from Eq. (4.53)], so that its local field fluctuates at a similar rate. This local field is then less effective in broadening the resonance line because only local fields that remain substantially constant over a period that is long compared with the Larmor precessional period and the duration τ_2 of the wave train are effective in shifting the resonance frequency of a given spin.

If the energy spectrum of an isolated ion is initially split, it cannot be counted as fully identical even with a similar neighbor. In this case an exchange interaction comparable in size to the splitting terms producing the structure may give rise to a broadening rather than a narrowing of the resonance line.

In most paramagnetic crystals the exchange interaction is not of the simple type arising from direct overlap of the electron wave functions that we discussed in connection with ferromagnetism but is rather of the superexchange or indirect exchange type. In the latter case interaction between each pair of paramagnetic ions may also depend on the arrangement of diamagnetic ions in between them; thus, the interaction is basically similar to that for magnetic dipole interaction. We can therefore expect anisotropic exchange to give rise also to a broadening of the resonance line.

6.3.3 Line Shapes

Computation of line shapes in principle can be carried out if the interaction constants are known for all pairs of spins in a crystal. In general, this restricts the computations to cases where the interaction is entirely due to magnetic dipole interaction between the spins, a situation that is rather rare if the magnetic moments are electronic rather than nuclear in origin. In fact, even if these conditions were satisfied, the mathematics involved in the calculation of the line shape would be prohibitively difficult. For this reason simple analytical forms for the line shape, though only approximately correct, can be very useful.

The two most commonly used line shape functions are *Lorentzian* and *Gaussian*. The Lorentzian shape factor is given by the expression

$$f_L(v) = \frac{1}{\pi} \frac{\Delta v}{(\Delta v)^2 + (v - v_0)^2} \tag{6.71}$$

where $\Delta v = 1/2\pi\tau_2$. For such a line shape the intensity falls to one-half the maximum value when the frequency deviates by $\pm \Delta v$ from the central frequency v_0. For the Gaussian line the shape factor is given by the Gaussian error function:

$$f_G(v) = \frac{1}{(2\pi \langle \Delta v^2 \rangle)^{1/2}} e^{-(v - v_0)^2/2\langle \Delta v^2 \rangle} \tag{6.72}$$

which is normalized and contains no arbitrary parameter, since it gives a mean *second moment* just equal to $\langle \Delta v^2 \rangle$, defined by the relation

$$\langle \Delta v^2 \rangle = \int (v - v_0)^2 f(v) \, dv \Big/ \int f(v) \, dv$$

It is interesting to note that χ'' given by Eq. (6.26) and $f_L(v)$ are of the same form, indicating that the Bloch resonance line at low signal levels is Lorentzian.

The line in such cases, where a characteristic time τ_2 can be attributed to the relaxation process involved, is said to be *homogeneously broadened*. The spins emit or absorb wave trains of finite length whose mean duration in time is τ_2 and whose probability distribution is of the form e^{-t/τ_2}; the Fourier transform of e^{-t/τ_2} is proportional to the Lorentzian function $f_L(\nu)$. The long tails of the Lorentzian line makes it essentially different from the Gaussian line for which the tails are very weak. Many observed lines are of the Lorentzian form over the region of observation. The tail far from resonance is usually difficult to follow because of the presence of noise.

In some other cases the Gaussian distribution gives a better fit to the experimentally observed lines. Whereas the Lorentzian function applied to cases where the line shape is due to the finite lifetimes of the levels participating in the transition, the Gaussian function results most often in absorption or emission from gaseous particles. The resulting radiation is Doppler-shifted owing to the velocity spread of the emitting particles. For a gas in thermal equilibrium the velocity distribution is Maxwellian and the resulting frequency distribution function is Gaussian.

For a normally paramagnetic salt it has been found that its linewidth due to spin-orbit interaction, inhomogeneous broadening, etc., is about 10^2–10^3 Oe. Since many fine or hyperfine splittings are much smaller, their resolution is greatly handicapped by such broad overall width. However, since the spin-spin interaction falls off rapidly with increasing interionic distance, considerable reduction in linewidth can be achieved by the use of "magnetically dilute" salts in which the majority of the paramagnetic ions are replaced by suitable diamagnetic ions in an isomorphic crystal. With sufficiently low concentration, broadening due to spin-spin interaction between paramagnetic ions can be reduced to a point at which it is less than that due to nuclear magnetic dipole moments that are not removed in the process of dilution. In crystals where the abundance of nuclear moments is low or zero, linewidths of a fraction of an oersted can be obtained.

Before concluding our discussion on spin-spin relaxation time, we may point out that spin-spin interaction is temperature-independent, giving rise to a line broadening that is itself independent of temperature when $2\mu_B H/kT \ll 1$. There is a small shift in the center of the line that increases with increasing $2\mu_B H/kT$, since orientation of neighboring dipoles parallel to the applied field $\mathbf{H}_0$ is statistically more probable than antiparallel orientations. When $2\mu_B H/kT \gg 1$, the dipoles are oriented parallel to $\mathbf{H}_0$ and the resonance line becomes sharp and appreciably shifted.

If the local field is due to other species of ions whose spin-lattice relaxation time is τ_1, the local field will be averaged out. This is a form of motional narrowing that can be observed in a compound that contains two paramagnetic species, one with an abnormally short τ_1 and the other with a long τ_2 (and therefore also a long τ_1) such as neodymium ethylsulfate containing a small fraction of gadolinium ions. At 90°K no resonance of the Nd^{3+} ions was observed due to their very small value of τ_1 ($<10^{-11}$ s) and consequent large broadening, while the Gd^{3+} resonance was observed with a linewidth substantially the same as in a diamagnetic host lattice such as lanthanum ethylsulfate. At 20°K, where the value of τ_1 for the Nd^{3+} ions is at least 10^{-8} s, the full effect of their magnetic dipolar field on the Gd^{3+} resonance spectrum is observed.

6.4 EXPERIMENTAL TECHNIQUES

Paramagnetic resonance can be observed by means of the microwave spectrometer shown in Fig. 6.3. In this arrangement we implicitly assume that the resonance frequency (but not the Q) of the sample cavity is not appreciably changed as the applied field H_0 is varied through resonance. If this is not the case, the sample should be placed in the reference cavity instead.

In the microwave (1000–100,000 MHz, say) spectrometer shown in Fig. 6.3, the sample is usually placed inside a rectangular cavity where the electric field is zero while the magnetic field is maximum. The reflected energy from the cavity whose magnitude is dependent on the static field applied to the sample is monitored by a directional coupler as shown. As the static field is varied through resonance at a given frequency, the absorption of the sample and of the cavity passes through a maximum. If the cavity is undercoupled, the reflected energy from the cavity increases with sample absorption. Since the reflected energy from the cavity under certain conditions, as discussed below, is proportional to the imaginary part of the diagonal component of the susceptibility

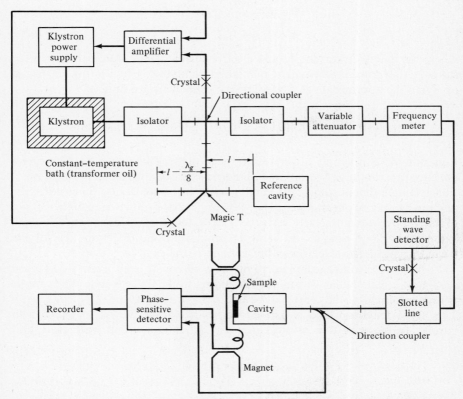

Figure 6.3 Schematic diagram of a microwave spectrometer for magnetic resonance measurements.

tensor, the reflection vs static field curve has the same shape as the χ'' vs H_0 curve desired.

When a paramagnetic sample is placed in a cavity, both the resonance frequency and the Q of the cavity are different from those of the empty cavity. As may be expected, to a first approximation, the change of the cavity resonance frequency is related to the real part of the susceptibility tensor components (since the insertion of the sample causes a redistribution of the electromagnetic fields in the cavity), and the change in Q is related to the imaginary part of the tensor components (since the sample increases the total loss of the cavity).

For the case of the empty cavity resonating at the frequency ω_0, Maxwell's equations are

$$\mathbf{\nabla} \times \mathbf{E}_0 = -j\omega_0\mu_0\mathbf{h}_0 \qquad (6.73)$$

$$\mathbf{\nabla} \times \mathbf{h}_0 = j\omega_0\epsilon_0\mathbf{E}_0 \qquad (6.74)$$

where $\mathbf{E}_0$ and $\mathbf{h}_0$ with time dependence $e^{j\omega_0 t}$ represent empty cavity fields. If a small sample of volume Δv is introduced into it, then Maxwell's equations for the sample-containing cavity become

$$\mathbf{\nabla} \times \mathbf{E} = -j\omega\mu_0\mathbf{h} - \mathbf{J}_m \qquad (6.75)$$

$$\mathbf{\nabla} \times \mathbf{h} = j\omega\epsilon_0\mathbf{E} + \mathbf{J}_e \qquad (6.76)$$

where the magnetic current $\mathbf{J}_m$ and electric current $\mathbf{J}_e$ are nonzero only at the sample. After some mathematical manipulation involving Eqs. (6.73)–(6.76) and applying the appropriate cavity boundary conditions, we find[17,18]

$$\omega - \omega_0 = j \int_{\Delta v} (\mathbf{J}_e \cdot \mathbf{E}_0^* + \mathbf{J}_m \cdot \mathbf{h}_0^*)\, d\tau \Big/ \int_v (\epsilon_0 \mathbf{E}_0^* \cdot \mathbf{E} + \mu_0 \mathbf{h}_0^* \cdot \mathbf{h})\, d\tau \qquad (6.77)$$

Equation (6.77) is simply a generalization of the well-known Bethe-Schwinger perturbation result for isotropic medium to the case of an anisotropic medium.[19] For a sample that is sufficiently small compared to the cavity dimensions, the fields in the cavity outside the sample (except perhaps in the immediate vicinity of the sample) are not greatly perturbed. Thus, to a first approximation, we can assume that $\mathbf{E} = \mathbf{E}_0$ and $\mathbf{h} = \mathbf{h}_0$ in the denominator of Eq. (6.77), and the volume v can be considered to be the entire volume of the cavity.

A suitable nondegenerated cavity can be constructed from a closed-off section of a rectangular waveguide resonating in the TE_{10n} mode. The sample is usually placed at the center of the end plate, where the electric field is zero. This static magnetic field is applied perpendicular to the r.f. magnetic field parallel to the wide dimension of the waveguide at the end wall. Note that

[17] J. O. Artman and P. Tannenwald, *J. Appl. Phys.* **26**, 1124 (1955).

[18] R. F. Soohoo, *Theory and Application of Ferrites*, Prentice-Hall, Englewood Cliffs, NJ, 1960, p. 260.

[19] H. A. Bethe and J. Schwinger, "Perturbation Theory for Cavities," National Defense Research Committee Contractors Rep. D1-117, Cornell University (March 4, 1943).

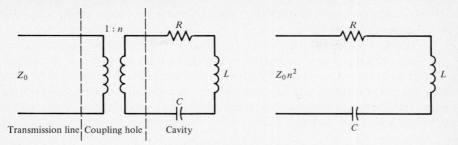

Figure 6.4 Equivalent circuit of a cavity coupled to a transmission line or waveguide. **(a)** The coupling hole represented by a $1:n$ ideal transformer. **(b)** $Z_0 n^2$ represents the transformed impedence of the waveguide.

$J_m = j\omega\mu_0\chi_p h$ where χ_p is the Polder susceptibility tensor given by[20,21]

$$\|\chi_p\| = \begin{bmatrix} \chi_p & -j\kappa_p & 0 \\ j\kappa_p & \chi_p & 0 \\ 0 & 0 & 0 \end{bmatrix} \tag{6.78}$$

with

$$\chi_p = \frac{(-\gamma H_0 + j/\tau_2)(-\gamma M_0)}{(-\gamma H_0 + j/\tau_2)^2 - \omega^2}$$

$$\kappa_p = \frac{\gamma M_0 \omega}{(-\gamma H_0 + j/\tau_2)^2 - \omega^2} \tag{6.79}$$

Carrying out the calculation indicated above for the rectangular cavity of volume v containing a sample of volume Δv, we find

$$\chi_p' = -\frac{v/\Delta v}{2[1 - (\lambda_0/2a)^2]}\left(\frac{\Delta\omega}{\omega_0}\right) \tag{6.80}$$

$$\chi_p'' = \frac{v/\Delta v}{2[1 - (\lambda_0/2a)^2]}\Delta\left(\frac{1}{2Q_u}\right) \tag{6.81}$$

where a is the width of the cavity, λ_0 the free-space wavelength, $\Delta\omega$ the frequency deviation from resonance, and ω_0 the resonance frequency of the cavity.

We shall now look into the actual measuring problem in somewhat more detail. Consider the equivalent circuit of a cavity fed by a transmission line of characteristic impedance Z_0 through a line transformer, as shown in Fig. 6.4; the value of n is dependent on the geometry of the cavity-coupling hole. Accordingly, we could define three separate cavity Qs: $Q_u = \omega_0 L/R$, $Q_e = \omega_0 L/n^2 Z_0$, and $Q_L = \omega_0 L/(R + n^2 Z_0)$ where Q_u, Q_e, and Q_L are the unloaded, external, and loaded Q of the cavity with series line loss neglected. Note that $1/Q_L = 1/Q_e + 1/Q_u$. If the cavity is undercoupled, $Q_u < Q_e$ and $R > n^2 Z_0$ so

---

[20]D. Polder, *Phil. Mag.* **40**, 99 (1949).

[21]If h at the sample is circularly rather than linearly polarized, $\chi_p + \kappa_p$ and $\chi_p - \kappa_p$ will be measured instead. Note that $\chi_p + \kappa_p$ and $\chi_p - \kappa_p$ correspond to χ_+', χ_+'' and χ_-', χ_-'' of Eqs. (6.25) and (6.26) and (6.29) and (6.30), respectively.

that at cavity resonance the VSWR is simply equal to $R/n^2 Z_0$. Thus, it follows that

$$\Delta\left(\frac{1}{Q_u}\right) = \frac{1}{Q_e} \Delta(\text{VSWR}) \tag{6.82}$$

Since the value of Q_e depends on the manner of coupling into the cavity, it is in general a constant during a particular experiment, so $\Delta(1/Q_u) = \Delta(1/Q_L)$. We find from Eqs. (6.81) and (6.82) that

$$\chi_p'' = C_1 \, \Delta(\text{VSWR}) \tag{6.83}$$

where the proportionality constant C_1 is given by

$$C_1 = \frac{v/\Delta v}{4Q_e[1 - (\lambda_0/2a)^2]} \tag{6.84}$$

According to Eq. (6.83), $\chi_p'' \propto \Delta(\text{VSWR})$. But VSWR is in turn related to the reflection coefficient ρ by the relation

$$\text{VSWR} = \frac{1 + |\rho|}{1 - |\rho|} \tag{6.85}$$

For small $|\rho|$ we see from Eq. (6.85) that $\Delta(\text{VSWR}) \propto \Delta|\rho|$, where $\Delta|\rho| = |\rho| - |\rho_0|$ and ρ is the reflection coefficient with the sample biased by a magnetic field. For maximum sensitivity ρ_0 should be zero; i.e., the cavity should be perfectly matched to the waveguide when the sample is biased off resonance by a large clamping field. However, if $\rho_0 = 0$, the reflected power may be very sensitive to mechanical vibrations of the spectrometer; as a compromise between sensitivity and stability, the VSWR of the empty cavity is usually made equal to about 2 by use of a coupling hold of appropriate size. For small samples $|\rho|$ is sufficiently small that the χ_p'' vs H_0 curve may be simulated by the $\Delta|\rho|$ vs H_0 curve. However, if $|\rho|$ is not sufficiently small, $\Delta(\text{VSWR})$ will not be proportional to $\Delta|\rho|$; then the linewidth will no longer correspond to the separation between half-power points of the $\Delta|\rho|$ vs H_0 curve, and ΔH must be measured by a more elaborate method.[22]

If dc detection is used, the reflected power due to finite $|\rho_0|$ may be balanced out simply by a dc voltage. For higher sensitivity a synchronous or phase-sensitive detection scheme[23] should be used. In this method the static field H_0 is modulated by a small-amplitude ac field with a frequency f_m ranging up to 150 kHz. Therefore, the reflected signal will contain an f_m component that can be detected by amplification. Using this method, the derivative of the absorption or the $d\chi_p''/dH_0$ curve rather than the absorption curve itself is obtained since the amplitude of the ac reflected signal is proportional to the slope of the $\Delta|\rho|$ vs H_0 curve. The advantage of this method lies in the fact that since the output of the synchronous detector is sensitive to the phase of the

[22]R. F. Soohoo, *Theory and Application of Ferrites*, p. 95.

[23]B. Chance et al., *Waveforms*, MIT Radiation Laboratory Series, No. 19, McGraw-Hill, New York, 1949, p. 515.

signal, the effective bandwidth of the device is extremely small with an accompanying reduction in noise. This is accomplished by feeding the detector with both the desired signal and a synchronizing signal from the source; these signals are usually connected to the grids of a twin-triode amplifier. If the time constant of the R-C grid coupling circuit is sufficiently long compared to the periods of the undesired signals, essentially only signals of frequency f emerge from the detector.

To stabilize the microwave frequency, isolators are used in the measuring system, as shown in Fig. 6.4. In addition, a microwave discriminator is also inserted to obtain high-frequency stability (~ 1 part in 10^6 long term and 1 part in 10^8 short term). Such a discriminator is simply a microwave analog of the ordinary r.f. discriminator.[24] The output from the two arms of the discriminator are equal when the Klystron output frequency is equal to the resonance frequency of the reference cavity. If these signals are applied to the inputs of a differential amplifier, there will be no output from the amplifier. However, if the frequency deviates from the desired value, the outputs from the discriminator arms will not be equal and the differential amplifier will have an amplified output proportional to their difference. If this signal is in turn applied to the reflector plate of the Klystron, the Klystron frequency will be changed until the signal frequency is equal to the reference value. (However, since the dc component of the amplifier output is positive with respect to ground while the reflector voltage is negative with respect to ground, a large bucking voltage of several hundred volts is also required.) Oscillations of the feedback loop should be avoided and typically a gain of some 30 dB is required for the differential amplifier.

After the χ_p'' vs H_0 curve for a given sample has been experimentally determined, it is a simple matter to determine ΔH by merely measuring the field separation between the half-power points at which χ'' is equal to one-half its value at resonance. If synchronous detection is used, the linewidth can be taken as the field separation between the maximum and minimum of the dispersion curve. For a curve of Lorentzian shape, the linewidth so measured is $1/\sqrt{3}$ times the true linewidth determined from the absorption curve.

6.5 MASERS

The study of masers and lasers at present constitute, to a large measure, the field of quantum electronics, which studies the quantum character of matter in the generation or amplification of microwave, infrared, or optical radiation. This rather exciting field derives its vitality from a number of related, yet seemingly distinct, disciplines. Thus, some knowledge of quantum and statistical mechanics, optics, and material chemistry, as well as some microwave know-how, will be required for the study of the material in this chapter. In our study of masers and lasers, we need not, of course, have a comprehensive knowledge

[24]F. E. Terman, *Radio and Electronic Engineering*, McGraw-Hill, New York, 1955, p. 696.

of all these fields. Indeed, as we shall demonstrate below, the body of knowledge derived from the aforementioned disciplines that is pertinent to the study of masers and lasers is not unduly large. Starting from rather fundamental considerations, we shall develop the pertinent concepts of quantum and statistical mechanics before applying them to a detailed discussion of maser action and its associated devices.

The word *maser* is an abbreviation for microwave amplification by stimulated emission of radiation; *laser* is an abbreviation for light amplification by stimulated emission of radiation. Unlike the case of ordinary microwave tubes, such as Klystrons and traveling wave tubes, whose operation can be understood in terms of classical mechanics, the behavior of masers and lasers can be completely analyzed only in terms of quantum and statistical mechanics. Usually, in these cases, we deal with the interaction of electromagnetic fields with bound aggregates of charges rather than with free charges as in the case of electron beams in microwave tubes.

It is appropriate at this point to inquire into the advantages of the maser and the laser over conventional amplifiers. The chief advantage of the maser is its extremely low noise figure. In contrast to the case of the conventional microwave amplifiers, which always contain shot noise from the electron beam, the only significant source of noise for the maser is that of spontaneous emission from the paramagnetic specimen providing the amplification. This type of noise is extremely small at microwave frequencies and at liquid helium temperatures. Since the noise figure of an amplifier chain is determined mainly by that of the leading amplifier, providing its gain is sufficiently high, the maser is an invaluable device for amplifying weak signals in radio astronomy when placed just behind the receiving antenna. Because of its narrow bandwidth and extreme frequency stability, the maser operating as an oscillator is a very accurate frequency or time standard (atomic clock).

The laser is capable of producing optical radiation that is powerful, monochromatic, coherent, and unidirectional. In contrast to this, emission from other light sources is usually neither monochromatic nor coherent. The high directivity of the laser output enables it to be focused onto a very small spot whose diameter is in the order of a wavelength of light or a few thousand angstroms. Such an intense localized beam could be used to study nonlinear effects of matter at optical frequencies or for surgical purposes. Modulated, a laser beam could become a communication medium.

6.5.1 Thermodynamic Equilibrium and Statistical Steady State

For a material that is at thermal equilibrium, we can deduce from statistical mechanics that the average number of particles N_i^e in a quantum state in which E_i is the atomic or molecular energy is[25]

$$N_i^e = Ce^{-E_i/kT} \tag{6.86}$$

[25]For the derivation of this Boltzmann factor, see, e.g., A. van der Ziel, *Solid State Physical Electronics*, 3rd ed., Prentice Hall, Englewood Cliffs, NJ, 1976, p. 44.

Furthermore, the total number of atoms or molecules N must be equal to the totality of N_i^e summed over all quantum states of the system:

$$N = \sum_i N_i^e = C \sum_i e^{-E_i/kT} \tag{6.87}$$

where we have used Eq. (6.86). Dividing Eq. (6.86) by Eq. (6.87) we find

$$N_i^e = Ne^{-E_i/kT} \bigg/ \sum_i e^{-E_i/kT} = \frac{N}{Z} e^{-E_i/kT} \tag{6.88}$$

where $Z = \sum_i e^{-E_i/kT}$ is known as the partition function of the system.

To recapitulate, if a system is at thermal equilibrium, the population of the various quantum states is given by Eq. (6.88). Conversely, if the system is not a thermal equilibrium, the population of the various states will not be as specified by Eq. (6.88). A particular state, called the *statical steady state*, is of special significance to our discussion of masers and lasers. In this state the mean properties of the macroscopic system are time-independent, but the microscopic state populations are not necessarily given by Eq. (6.88). An example of such a state is a macroscopic system in contact with a thermal reservoir and subjected to a steady, monochromatic radiation field of frequency ω near some resonance frequency of the constituent microscopic system.

According to Eq. (6.88), the population of the various quantum states decreases with increasing energy; i.e., $N_j^e < N_i^e$ if $E_j > E_i$. It turns out that, even in the presence of an external electromagnetic field, the sample containing these energy levels can only absorb energy from the field so long as the above inequalities are not disturbed. However, if population inversion occurs, i.e., $N_j^e > N_i^e$, the sample can be made to emit rather than to absorb radiation. Indeed, the operation of masers and lasers are dependent on just such population inversion. Methods to achieve this inversion, called *pumping methods*, will be studied in some detail in this chapter.

6.5.2 Population Inversion

In the last section we indicated that to obtain maser action, a population inversion is needed. Here we shall first show that, in the absence of a population inversion, energy will always be absorbed by the microscopic system from the external electromagnetic field. Then we shall describe several means for achieving population inversion.

To illustrate our point, let us consider the simple problem of an isolated electron spin in a static magnetic field.

Our analysis in Section 6.3.1 in connection with the spin-lattice relaxation time showed that the rate of energy absorption is given by the expression

$$\frac{dE}{dt} = (N_1 - N_2)C\hbar\omega = nC\hbar\omega \tag{6.67}$$

where $N_1 C$ is the number of electron spins per unit time that go from the lower energy state to the higher one, $N_2 C$ is the number of electron spins per unit time that go in the reverse direction, and $\hbar\omega$ is the energy of a photon corresponding to a frequency ω. It follows that $N_1 C\hbar\omega$ corresponds to energy absorption while $N_2 C\hbar\omega$ corresponds to energy emission. Thus, a positive dE/dt represents net energy absorption while a negative dE/dt represents a net energy emission. Maser action corresponds to energy emission, which according to Eq. (6.67) requires that $N_2 > N_1$ or, equivalently, $n < 0$. However, according to Eq. (6.66), $n > 0$, since n_0, C, τ_1 are all positive. Consequently, not only is $N_2^e < N_1^e$, a thermal equilibrium according to Eq. (6.88), but $N_2 < N_1$ even at statistical steady state in the presence of an alternating magnetic field. Therefore, if maser action is to be obtained, some means of inverting the population so that $N_2 > N_1$ must be used. Methods for achieving population inversion will be discussed in detail next.

6.5.3 Inversion Methods

For a two-level system *inversion* and *amplification* are usually spatially separated. For example, in the ammonia maser the two processes are spatially separated in that the NH_3 molecules are first passed through a velocity selector to accomplish inversion before they enter the cavity to perform amplification. The velocity selector is so constructed that the ammonia molecules belonging to the lower vibrational inversion energy state are deflected out of the vapor beam by an inhomogeneous electric field. The remaining molecules of the upper state are in turn directed through a cavity. The electromagnetic field in this cavity induces transitions of the molecules from the upper to the lower state, emitting energy in the transition processes. The electromagnetic field associated with this emitted energy is in phase with the cavity field that induces the transition, thereby amplifying it.

  For a three-level maser population inversion could be obtained without spatial separation of the inversion and amplification processes as in the two-level maser. How this process is accomplished can best be understood by reference to Fig. 6.5. Consider first Fig. 6.5(a). Particles are induced to go from energy level E_1 to E_3 by a pump that is relatively strong compared to the amplified signal. If the relaxation time between E_3 and E_2 is relatively long compared to that between E_2 and E_1, a population inversion may develop between levels

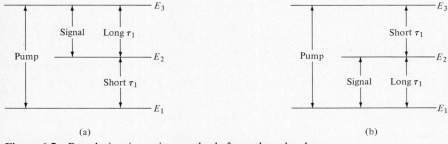

(a)   (b)

Figure 6.5 Population inversion methods for a three-level maser.

3 and 2. This is so because a particle at level E_2 will quickly relax to level E_1, thus helping to deplete E_2, whereas a particle in E_3 will take a relatively longer time to relax toward E_2, adding to the population excess of E_3. We shall now proceed to prove these statements mathematically.

First, we note from the state equation (6.64) that if the pump is sufficiently strong that C_{12} and C_{21} terms dominate, $N_1 \simeq N_2$ at steady state. This state of affairs is known as saturation pumping. Note, however, that for Fig. 6.5, the pump is connected between levels 1 and 3, so that instead of the $N_1 = N_2$ referred to above in connection with Eq. (6.64), we should have $N_1 \simeq N_3$. If the pumping action does not disturb level 2, the total population of levels 1 and 3 must be independent of time and indeed should be equal to that at thermal equilibrium. Thus,

$$N_1 + N_3 = N_1^e + N_3^e \tag{6.89}$$

Or, recalling that $N_1 \simeq N_3$, we find from Eq. (6.89) that

$$N_1 \simeq N_3 \simeq \tfrac{1}{2}(N_1^e + N_3^e) \tag{6.90}$$

To demonstrate population inversion, we must show that

$$N_3 > N_2 \simeq N_2^e \tag{6.91}$$

where we again assume that level 2 is not sufficiently disturbed. Combining Eqs. (6.90) and (6.91), we must show equivalently that

$$\frac{N_1^e}{N_2^e} + \frac{N_3^e}{N_2^e} > 2 \tag{6.92}$$

Substituting the expressions for N_1^e, N_2^2, etc., by means of Eq. (6.88), Eq. (6.92) becomes

$$e^{(E_2 - E_1)/kT} + e^{(E_2 - E_3)kT} > 2 \tag{6.93}$$

Now let $E_2 - E_1 = (E_3 - E_2) + \delta$; then Eq. (6.93) reads

$$e^{[(E_3 - E_2) + \delta]/kT} + e^{-(E_3 - E_2)/kT} > 2 \tag{6.94}$$

An examination of Eq. (6.94) shows that inequality (6.94) is satisfied if δ is positive for all values of $E_3 - E_2$. Equivalently,

$$(E_3 - E_1) > 2(E_3 - E_2)$$

that is, the pump frequency must be more than twice the signal frequency if population inversion were to be obtained. This inequality will be substantiated in Section 6.5.4 by a detailed calculation of the emissive power of the three-level maser.

Another three-level maser scheme is shown in Fig. 6.5(b). Here the signal is applied across levels 2 and 1 instead of between levels 3 and 2 as in Fig. 6.5(a). Correspondingly, the relaxation time between E_3 and E_2 will be shorter than that between E_2 and E_1, while the converse is true for the configuration of Fig. 6.5(a). Thus, referring to Fig. 6.5(b), a particle in E_3 will relax quickly to E_2 because of the short relaxation time between them. On the other hand, since the relaxation time between E_2 and E_1 is long, it is possible to maintain a population inversion between E_2 and E_1.

6.5.4 The Three-Level Maser

Refer now to the energy diagram of a three-level maser depicted in Fig. 6.5(a). The rate equations for this system can be written immediately by analogy with Eq. (6.64) for the two-level case:

$$
\begin{aligned}
\dot{N}_3 &= (P_{13}N_1 - P_{31}N_3) + (P_{23}N_2 - P_{32}N_3) \\
&\quad + C_p(N_1 - N_3) + C_s(N_2 - N_3) \\
\dot{N}_2 &= (P_{32}N_3 - P_{23}N_2) + (P_{12}N_1 - P_{21}N_2) + C_s(N_3 - N_2) \\
\dot{N}_1 &= -\dot{N}_2 - \dot{N}_3
\end{aligned}
\tag{6.95}
$$

Where $C_p = C_{13} = C_{31}$ is the induced transition probability between pump levels 1 and 3, and $C_s = C_{23} = C_{32}$ is that between signal levels 2 and 3.

Under statistical steady-state conditions, $\dot{N}_1 = \dot{N}_2 = \dot{N}_3 = 0$. Solving Eq. (6.95), we find the normalized population difference between levels 3 and 2:

$$
\frac{N_3 - N_2}{N_1} = -\frac{C_1 + (P_{12} - P_{21} + P_{32} - P_{23})C_p}{(P_{31} + C_p)(P_{23} + P_{21} + C_s) + P_{21}(P_{32} + C_s)}
\tag{6.96}
$$

where

$$
C_1 = P_{12}(P_{31} - P_{23}) - P_{13}(P_{21} + P_{23}) + P_{32}(P_{12} + P_{13})
\tag{6.97}
$$

In a similar manner, we also find

$$
\frac{N_3 + N_2}{N_1} = \frac{C_2 + (P_{12} + P_{21} + P_{32} + P_{23} + 2C_s)C_p + 2(P_{12} + P_{13})C_s}{(P_{31} + C_p)(P_{23} + P_{21} + C_s) + P_{21}(P_{32} + C_s)}
\tag{6.98}
$$

where

$$
C_2 = P_{12}(P_{31} + P_{23}) + P_{13}(P_{21} + P_{23}) + P_{32}(P_{12} + P_{13})
\tag{6.99}
$$

Furthermore, since the total number of particles are conserved ($N_1 + N_2 + N_3 = N$), it follows that

$$
N_1 = \frac{N}{1 + (N_3 + N_2)/N_1}
\tag{6.100}
$$

These equations can be simpiified considerably if we assume that the pump is sufficiently strong that C_p is much greater than either C_s or the Ps. In this case Eqs. (6.96) and (6.98) become

$$
\frac{N_3 - N_2}{N_1} \simeq -\frac{P_{12} - P_{21} + P_{32} - P_{23}}{P_{23} + P_{21} + C_s}
\tag{6.101}
$$

$$
\frac{N_3 + N_2}{N_1} = \frac{P_{12} + P_{21} + P_{32} + P_{23} + 2C_s}{P_{23} + P_{21} + C_s}
\tag{6.102}
$$

If we further stipulate that hv_{21}, $hv_{32} \ll kT$ as may be the case with masers, Eqs. (6.101) and (6.102) may be further simplified by noting that in this case

$$P_{21} = P_{12}e^{hv_{21}/kT} \simeq P_{12}\left(1 + \frac{hv_{21}}{kT}\right)$$
$$P_{32} = P_{23}e^{hv_{32}/kT} \simeq P_{23}\left(1 + \frac{hv_{32}}{kT}\right) \tag{6.103}$$

where we have used Eq. (6.53) and noted that $hv_{21} = E_2 - E_1$ and $hv_{32} = E_3 - E_2$. Using Eq. (6.103) and the statement of particle conservation ($N_1 + N_2 + N_3 = N$), we find from Eqs. (6.101) and (6.102)

$$N_3 - N_2 \simeq \frac{hN}{3kT} \frac{P_{12}v_{21} - P_{23}v_{32}}{P_{12} + P_{23} + C_s} \tag{6.104}$$

We have shown previously that in order to achieve stimulated emission there must first be a population inversion, i.e., $N_3 > N_2$. Since all quantities appearing in Eq. (6.104) are positive, it follows that population inversion is possible if and only if

$$P_{12}v_{21} > P_{23}v_{32} \tag{6.105}$$

Now, by definition,

$$v_{21} = v_{31} - v_{32} \tag{6.106}$$

Combining Eqs. (6.105) and (6.106), we find

$$v_{31} > \left(1 + \frac{P_{23}}{P_{12}}\right)v_{32} \tag{6.107}$$

according to Eq. (6.59), P_{23}/P_{12} is, to a first approximation, nearly equal to $(\tau_1)_{12}/(\tau_1)_{23}$ where $(\tau_1)_{12}$ and $(\tau_1)_{23}$ are, respectively, relaxation times between levels 1 and 2 and between levels 2 and 3. Since $(\tau_1)_{12}$ must be smaller than $(\tau_1)_{23}$ according to the discussion in Section 6.5.3, Eq. (6.107) yields the inequality

$$v_{31} > Kv_{32}$$

where the constant K for a given medium has a value between 1 and 2. However, if the relaxation times are equal, Eq. (6.107) shows that $v_{31} > 2v_{32}$, as stated in connection with Eq. (6.94), which was derived by assuming that level 2 is not disturbed. Thus, the three-level maser requires a source of saturating power at a frequency approximately double that of the signal frequency.

The emitted power is equal to the total number of emitted photons per second, $C_s(N_3 - N_2)$, times the energy of each photon hv_{32}. Thus,

$$P_e = C_s(N_3 - N_2)hv_{32} = \frac{Nh^2v_{32}C_s}{3kT} \frac{P_{12}v_{21} - P_{23}v_{32}}{P_{12} + P_{23} + C_s} \tag{6.108}$$

Let us now examine the various factors involved in the above expression for the emitted power. First, we have already discussed the relationship between the pump and signal frequencies. Second, we note that P_e is directly proportional to N. It turns out, however, that as N increases, P_e reaches some maximum value and thereupon decreases as N is further increased. There is a finite optimum value of N in Eq. (6.108) because the various particles are not isolated from each other as assumed in its derivation. As the number of paramagnetic ions in the diamagnetic salt N increases, the mutual interaction between them increases, and this eventually leads to a decrease in P_e with further increase in paramagnetic ion concentration.

It may also be noted from Eq. (6.108) that P_e is inversely proportional to T, indicating the desirability of using low temperatures. Of course, the (2, 3) signal transitions must be allowed.

Finally, we note that if $C_s \ll P_{12} + P_{23}$, then the emitted power is proportional to the induced signal transition probability C_s itself, signifying linear amplification. If the signal transition is saturated, i.e., if C_s cannot be neglected compared to $P_{12} + P_{23}$, then P_e is no longer proportional to C_s, signifying nonlinear amplification.

The three-level maser was first proposed by Bloembergen.[26,27] Later, Schovil, Feher, and Seidel constructed the first experimental device.[28] They used the spin resonance of the Gd^{4+} paramagnetic ion[29] of diluted gadolinium ethylsulfate at $1.2°K$. The ground state of Gd^{4+} ion is ^8S, indicating that there are eight electrons with a total orbital angular momentum L equal to zero and a total spin angular momentum S equal to $\frac{7}{2}$. In the absence of the crystalline field, the ground state is split into eight ($= 2S + 1$) equally spaced energy levels characterized by the quantum number S_z having half-integral values, $-\frac{7}{2}$, $-\frac{5}{2}, \ldots, +\frac{5}{2}, +\frac{7}{2}$ when a static magnetic field is applied in the direction z. The interaction between the ionic spin and the crystalline field of the solid may cause the energy levels to be nonequidistant, as exemplified by the actual Gd^{4+} energy level diagram of Fig. 6.6. As shown in the figure, a signal of 9 GHz is applied between the $S = -\frac{5}{2}$ and $S = -\frac{3}{2}$ states, and a pump of 17.5 GHz is applied across the $S = -\frac{5}{2}$ and $S = -\frac{1}{2}$ states.

In the maser action we are concerned with the resonance of the paramagnetic ion imbedded in a diamagnetic crystalline lattice. Two complications already mentioned above deserve further attention. First, in paramagnetic ions several electronic spins may be strongly coupled together in such a way that the total spin is larger than $\frac{1}{2}$. In this case the number of possible energy levels is equal to $2S + 1$ where S is equal to the total spin in a given direction when the individual spins are all parallel. Thus, for the Gd^{4+} ion mentioned, $S = \frac{7}{2}$ since there are seven electrons giving rise to the effective spin. It follows that there are eight energy levels, since $2S + 1 = 8$. The second complication arises because of the crystalline field. When a paramagnetic ion is imbedded in a crystalline lattice, the neighboring ions produce an electric field that acts upon it

[26]N. Bloembergen, *Phys. Rev.* **104** 324 (1956).

[27]This is the solid-state version of the ammonia maser first proposed by J. P. Gordon, H. J. Zeiger, and C. H. Townes, *Phys. Rev.* **99**, 1264 (1955).

[28]H. E. D. Scovil, G. Feher, and H. Seidel, *Phys. Rev.* **105**, 762 (1957).

[29]An ion is paramagnetic if its effective spin is nonzero.

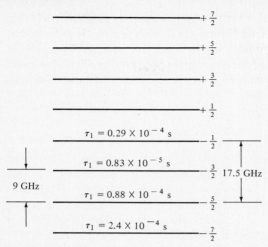

Figure 6.6 Energy levels for the $6d^{4+}$ ion.

and distorts the motion of its paramagnetic electrons. Since the electron orbits in general differ among the states with different components of S along the magnetic field, the energy levels of the corresponding electron states would be unequally perturbed. Thus, the energy of interaction between the electron spins and the magnetic field is no longer linear in H, as is the case with the free electron of Fig. 5.4. A typical example of crystal field distortion is given in Fig. 6.7 for a ruby crystal that contains a small amount of chromium in aluminum oxide (AL_2O_3). The solid lines indicate the energy levels in the absence of a crystalline field, while the dashed lines represent the same levels in the presence of a crystalline field. Note that here the crystal field has split the fourfold degenerate levels in zero field into a pair of twofold degenerate levels while the application of a magnetic field further changes the levels in a nonlinear manner. Furthermore, the energies, except for the case of cubic crystals, are dependent on the direction of the magnetic field relative to the crystalline axes. Note that the frequency to be amplified could be changed by changing the magnetic field; of course, the pump frequency must be adjusted accordingly.

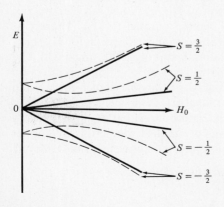

Figure 6.7 Cr^{3+} ions in ruby in a static field.

It is significant to note that since the splittings in a magnetic field are purely Zeeman in character, the energy levels are equally spaced. It follows that it is impossible to pump and amplify only between the selected three levels. However, because of the presence of the superimposed crystalline field, the levels corresponding to different spin states are, in general, unevenly spaced, thus rendering maser level selection possible.

6.5.5 Two-Level Maser

The first successful demonstration of maser action, by Gordon, Zeiger, and Townes, used the so-called inversion levels of ammonia gas molecules.[30] In this type of maser a molecular beam of ammonia is passed through an inhomogeneous electrostatic field that effectively focuses the molecules in the upper energy state and defocuses those in the lower energy state. The focused beam composed of upper-state molecules is then passed through a microwave cavity that has a mode with a resonance frequency equal to the inversion frequency of the ammonia molecule. If the cavity is excited to oscillate at this particular mode by a microwave signal in the presence of the molecular beam, the ammonia molecules can be induced to go from the upper state to the lower state. Since the radiation from these molecules are in phase with the microwave signal inducing it, the signal is thereby amplified. For a sufficiently intense beam, oscillation may also occur. There are no molecules in the lower energy state as the beam enters the microwave cavity so that a population inversion has been accomplished in this case by preselection, rather than by pumping as in the case of the three-level maser.

Population inversion can also be accomplished by a process known as *adiabatic rapid passage*. Consider the case of a rotating magnetic field **h** of frequency ω, perpendicular to a static field $\mathbf{H}_0$. Far below resonance, the magnetization **M** is nearly parallel to the effective field $\mathbf{H}_{\text{eff}}$ with $|\mathbf{H}_{\text{eff}}|$ given by

$$|\mathbf{H}_{\text{eff}}| = \sqrt{h^2 + \left(\frac{\omega}{|\gamma|} - H_0\right)^2} \tag{5.19}$$

where γ is the gyromagnetic ratio and $\mathbf{H}_{\text{eff}}$ is an effective field as seen by an observer in a coordinate system in which **h** is stationary. As resonance is approached, both the magnitude and direction of $\mathbf{H}_{\text{eff}}$ changes, but if resonance is approached sufficiently slowly, **M** will remain parallel to $\mathbf{H}_{\text{eff}}$. Exactly at resonance, i.e., at $\omega/|\gamma| = H_0$, $\mathbf{H}_{\text{eff}} = \mathbf{h}$ according to the expression for $\mathbf{H}_{\text{eff}}$ [Eq. (5.19)] so that the magnetization will lie along **h** making a 90° angle with $\mathbf{H}_0$. If we vary $\mathbf{H}_0$ slowly through resonance, **M** will end up pointing in a direction opposite to $\mathbf{H}_0$. This technique for inverting **M** so that the higher-energy spin state is relatively more populated is called *adiabatic inversion*.

The discussion above is based on the assumption that H_0 varies slowly through resonance. It would be instructive to establish the permissible upper limit to the rate of variation of H_0 if adiabatic inversion were to be achieved.

[30]Gordon, Zeiger, and Townes, *Phys. Rev.* **99**, 1264 (1955).

A rapidly changing H_0 will have oscillatory components that introduce transitions between the atomic energy levels.

To begin with, let us assume that the direction rather than the magnitude of $\mathbf{H}_0$ is varied; the resultant effect is similar to a variation of the magnitude of $\mathbf{H}_0$ as $\mathbf{H}_{eff}$ changes in direction with changing magnitude of $\mathbf{H}_0$. Assume further that this change consists of a rotation about the x axis with an angular velocity ω_1 as depicted in Fig. 6.8. As in the case of superimposed orthogonal static and r.f. magnetic fields, it is mathematically expedient to change from the laboratory frame to one rotating with a velocity equal to ω_1. In the rotating system, aside from a constant field $\mathbf{H}_0$, there will also be an apparent field $-\omega_1/|\gamma|$ at right angles to $\mathbf{H}_0$, as shown in Fig. 6.8. Thus, the effective field $\mathbf{H}_{eff}$ in the rotating frame is the vector sum of $\mathbf{H}_0$ and $-\omega_1/|\gamma|$. If $\mathbf{M}$ was initially parallel to $\mathbf{H}_0$, then it will start to precess about $\mathbf{H}_{eff}$ as $\mathbf{H}_0$ is rotated. Hence, $\mathbf{M}$ will remain closely parallel to $\mathbf{H}_0$ only if the angle θ_1 between $\mathbf{H}_0$ and $\mathbf{H}_{eff}$ is very small. Since

$$\tan \theta_1 = \frac{\omega_1}{|\gamma| H_0} \tag{6.109}$$

according to Fig. 6.8, we conclude that for $\mathbf{M}$ to remain parallel to $\mathbf{H}_0$, the inequality

$$\omega_1 \ll |\gamma| H_0 \tag{6.110}$$

must be satisfied. This inequality says that the frequency of rotation of the field $\mathbf{H}_0$ must be much smaller than the Larmor frequency of precession in the field $\mathbf{H}_0$ to ensure that $\mathbf{M}$ and $\mathbf{H}_0$ are parallel. Inasmuch as paramagnetic resonance frequencies typically lie in the microwave band (10^{10}–10^{12} Hz, say) while the modulation of $\mathbf{H}_0$ is seldom used at a frequency much above 10^5 Hz, the adiabatic condition is easily satisfied in practice.

Let us now apply the adiabatic condition (6.110) to the case of paramagnetic resonance. From Eq. (5.19) and Fig. 5.3(b), we have, for the rotational

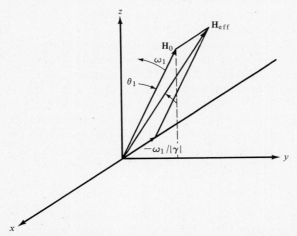

Figure 6.8 Effective field $\mathbf{H}_{eff}$ due to rotation of the applied field $\mathbf{H}_0$ about the x axis.

frame,

$$\mathbf{H}_{eff} = \mathbf{H}' + \mathbf{h} \tag{6.111}$$

where $\mathbf{H}' = \hat{\mathbf{z}}(H_0 - \omega/|\gamma|)$. If $H_0 > \omega/|\gamma|$, $\mathbf{M}$ is parallel to $\mathbf{H}_0$ initially, i.e., with $\theta = 0$ at $t = 0$. As H_0 is varied slowly through resonance, $\mathbf{M}$ follows $\mathbf{H}_{eff}$ and θ changes from 0 to π, as indicated in Fig. 5.3(a). On the other hand, if $H_0 < \omega/|\gamma|$, then $\mathbf{M}$ is initially antiparallel to $\mathbf{H}_0$, or $\theta = \pi$ at $t = 0$. As H_0 is varied slowly through resonance, $\mathbf{M}$ will again follow $\mathbf{H}_{eff}$ but θ will now change from π to 0.

For the paramagnetic resonance condition under consideration, ω_1 of Eq. (6.110) is analogous to $d\theta/dt$. Since $d/dt(\cot \theta) = -\csc^2 \theta (d\theta/dt)$ we have

$$\frac{d\theta}{dt} = -\sin^2 \theta \frac{d\theta}{dt}(\cot \theta) \tag{6.112}$$

Noting that $\sin \theta = h/H_{eff}$ and $\cot \theta = H'/h$, we can rewrite Eq. (6.112) to read

$$\frac{d\theta}{dt} = -\left(\frac{h}{H_{eff}}\right)^2 \frac{d(H_0/h)}{dt} \tag{6.113}$$

where we have made use of the definition $H' = H_0 - \omega/|\gamma|$ and the constancy of H_0. Since $H_{eff} \geq h$, the maximum value of $(h/H_{eff})^2$ is clearly unity. Furthermore, the rotating field H_0 of Eq. (6.110) corresponds to h in this case. It therefore follows from Eqs. (6.110) and (6.113) and the correspondence of ω_1 and $d\theta/dt$ that the adiabatic condition in the case of paramagnetic resonance is given by

$$\frac{1}{h}\left|\frac{dH_0}{dt}\right| \ll |\gamma|h \tag{6.114}$$

In other words this means that any change in H_0 in a time of the order $1/|\gamma|h$ must be much smaller than h itself. If $h = 1$ Oe, $\gamma = -2\pi(2.8) \times 10^6$ rad/s Oe, as for free electrons, then $|dH_0/dt|$ should be much less than 1 Oe/6×10^{-8} s. In other words H_0 must be varied in such a way that a change of 1 Oe takes place in a time that is long compared with 6×10^{-8} s.

In our discussion thus far it has been necessary to determine $\mathbf{M}$ by reference to the initial conditions, assuming the system was initially in thermal equilibrium. Such equilibrium is attained by means of relaxation processes that require a finite time. If the sweep of H_0 through resonance takes a time that is long compared with the spin-lattice relaxation time τ_1, reversal of $\mathbf{M}$ will not occur as relaxation will tend to restore $\mathbf{M}$ to being parallel to $\mathbf{H}_0$. It follows that there is a lower limit to the sweep rate through resonance as well as an upper limit given by Eq. (6.110). This lower limit, stated mathematically, is

$$\frac{1}{h}\left|\frac{dH_0}{dt}\right| \gg \frac{1}{\tau_1} \tag{6.115}$$

where we have noted from Eq. (6.114) that the rate of change of H_0 is $(1/h)(dH_0/dt)$. When $1/\tau_1$ is much smaller than $|\gamma|h$, it is clearly possible to satisfy both conditions. The term *adiabatic rapid passage* is used when the inequalities

$$\frac{1}{h}\left|\frac{dH_0}{dt}\right| \ll |\gamma|H_1 \gg \frac{1}{\tau_1} \tag{6.116}$$

are satisfied.

6.5.6 Noise

Since low noise is the most outstanding feature of a maser, it is appropriate to examine the subject of noise further. To begin with, the noise power inherent in a maser of bandwidth Δv is given by

$$P_n = 4K|T_s|\Delta v \tag{6.117}$$

This expression is the same as for a conventional system except the effective temperature T_s may be negative. Thus, P_n can be very low if T_s is numerically small, even when negative; Δv is small for a maser as the energy levels between which amplification occurs are usually very sharp in a dilute paramagnet.

The definition of spin temperature follows naturally from Boltzmann statistics. To begin with, in thermodynamic equilibrium

$$\frac{N_j^e}{N_i^e} = e^{-hv_{ji}/kT} \tag{6.118}$$

where we have used Eq. (6.88) and noted that $E_j - E_i = hv_{ji}$. In this case we have $N_j^e \le N_i^e$ at any ambient temperature T (the temperature of the helium dewar that contains the paramagnetic specimen, for example). If the spin system is not at thermodynamic equilibrium, we may still, in analogy with Eq. (6.118), write

$$\frac{N_j}{N_i} = e^{-hv_{ji}/kT_s} \tag{6.119}$$

Here N_j and N_i are populations of levels j and i, respectively. If the spin temperature T_s equals the ambient temperature T, then N_j and N_i clearly equal, respectively, the thermal equilibrium values N_j^e and N_i^e. In this context we can say that the spin system is at thermal equilibrium if the spin temperature is equal to the ambient temperature.

We have shown that at statistical steady state, $N_j > N_i$ owing to population inversion. It follows from Eq. (6.119) that T_s is negative in this case, as alluded to above.

6.5.7 Maser Structures

For a three-level maser both the pump and the signal must be coupled to the paramagnetic material. In order to maximize the magnetic interaction between the pump and signal and the material, it is desirable to place the material at a position of a high-Q cavity at which the r.f. magnetic field is maximum. Of course, if the material dielectric losses are sufficiently low at microwave frequencies, as is the case with ruby, it will be permissible to fill the entire cavity with the maser material since the cavity field is uniform in time phase. Furthermore, a maser material such as rutile can be used as a dielectric cavity and metal walls need not be used. Normally, the pump and signal are coupled to the cavity by separate coupling holes, as schematically shown in Fig. 6.9.

To prevent the maser from amplifying the noise from the next stage, i.e., amplifier A_1, a ferrite circulator must be used in the manner indicated in Fig. 6.9. The construction of the circulator is such that power entering the signal terminal of the circulator will emerge from the maser terminal while the amplified signal will return from the maser via the same maser-to-circulator waveguide to emerge from the amplifier terminal to enter the next stage, A_1. Any reflection from A_1 will go into the matched load and be dissipated. In order to maximize the maser gain bandwidth product and to minimize the noise level, it is frequently necessary to enclose not only the cavity but also the circulator in the helium dewar as ferrite losses are lower at lower temperatures. The magnetic field appropriate for a given combination of pump and signal frequencies can be adjusted by changing the current I in the electromagnet. Alternatively,

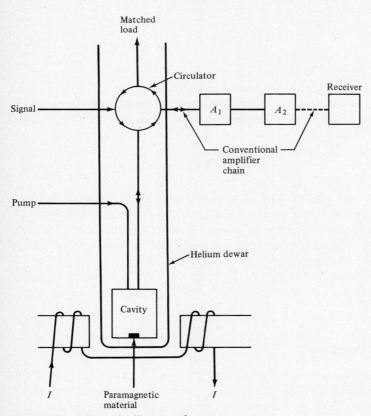

Figure 6.9 Schematic diagram of a maser.

a superconducting solenoid immersed inside the dewar may be conveniently used to provide the required magnetic field.

Since the field in a resonant cavity is enhanced by a factor of Q at resonance as compared to the field off resonance, the cavity maser has the advantage of a large field-material interaction in a reasonably small volume. On the other hand the high-Q cavity also limits the usable bandwidth of the maser. This situation can be alleviated by putting the material in a slow-wave rather than a cavity structure. In this device, known as the *traveling-wave maser*,[31] the electromagnetic field is weakly coupled to the structure as the r.f. field is no longer enhanced by a high Q, as in the case of the cavity structure. Nevertheless, high gain is obtained because the interaction takes place for an extended time and distance while the signal travels through the material placed in a slow-wave structure. The bandwidth is large compared to that of cavity maser as the slow-wave structure is a nonresonant device. To avoid possible oscillations, unilateral gain in the direction of wave propagation must be provided for, either by building ferrite isolation into the structure or by using maser materials of high paramagnetic concentration.

Slow-wave structures are used for a traveling-wave maser solely to reduce the physical length of the waveguide required to obtain useful gain. This can be understood on the basis of the following relation between the guide wavelength λ_g and the phase velocity v_p:

$$\lambda_g = \frac{v_p}{v} \tag{6.120}$$

where v is the frequency. Thus, if $v_p = 0.1c$ where c is the velocity of light, the slow-wave structure need only be one-tenth as long as a corresponding normal waveguide operated far above cutoff for the same gain.

From the energy conservation standpoint, the signal in a traveling-wave maser will be amplified if the energy gain by the signal in its interaction with the paramagnetic material exceeds the losses due to the slow-wave structure, material dielectric losses, and possibly ferrite losses. Similarly, for the cavity maser, the signal is amplified if the gain exceeds all losses in the structure. As with all other types of amplifiers, oscillation could be achieved if the gain is sufficiently large. An example of this is supplied by the ammonia maser, which acts as an oscillator at sufficiently high gas beam density so that it can be a very accurate frequency standard or atomic clock.

PROBLEMS

6.1 Sketch susceptibility curves similar to those shown in Fig. 6.1 for the negative circularly polarized case.

6.2 The steady-state solutions for Eqs. (6.34)–(6.36) are given by Eqs. (6.37)–(6.39). Find the corresponding transient solutions.

[31] R. W. DeGrasse, E. O. Schulz-DuBois, and H. E. D. Scovil, *Bell Syst. Tech. J.* **38**, 305 (1959).

6.3 **(a)** Sketch and compare the Lorentzian and Gaussian shape factors given by Eqs. (6.71) and (6.72).

(b) Find the expression for χ' from χ'' given by Eq. (6.72) for the Gaussian line. [Hint: Use the Kronig-Kramer relation.]

6.4 Referring to Fig. 6.3, sketch the power reflected from the cavity vs H_0 for the following three cases:

(a) Sample cavity undercoupled

(b) Sample cavity critically coupled

(c) Sample cavity overcoupled

6.5 Show under what experimental conditions only f_m (actually dc component of the rectified signal of frequency f_m) emerges from the phase-sensitive detector.

6.6 Show that if the χ''_+ vs H_0 curve is Lorentzian, the field separation between the maximum and minimum of $\partial \chi''/\partial H_0$ vs H_0 curve is $1/\sqrt{3}$ times the linewidth measured from the χ''_+ vs H_0 curve.

6.7 For a laser, in contrast to a maser, $h\nu_{21}$, $h\nu_{32}$ is not necessarily less than kT. Derive the population inversion expression [analogous to Eq. (6.104) for a maser] for a laser.

Chapter 7

Ferromagnetic and Antiferromagnetic Resonance

In Chapter 6 we discussed the paramagnetic resonance phenomenon associated with weakly interacting electron spins. In this chapter we shall examine the cases of ferromagnetic and antiferromagnetic resonance associated with strongly interacting electron spins. In ferromagnetic and antiferromagnetic materials these spins are coupled by extremely strong exchange forces. As a consequence of this strong coupling, the spins behave in many cases as a cooperative assembly rather than as more or less isolated spins. When the spins precess in unison at ferromagnetic resonance, the phenomenon is referred to as *uniform precession*. If the precession is nonuniform, the magnetization of the ferromagnet can be expressed as a summation of "spin waves." Both the uniform precession and spin waves will be discussed in some detail in this chapter.

Since practically all ferromagnetic materials are electrical conductors, their use in microwave devices is limited; at these frequencies their losses are prohibitively high unless they are situated at positions of zero electric field. For this reason applications of ferromagnetic materials at microwave frequencies use specimen in the form of thin films. The behavior of such films as well as that of bulk materials at microwave frequencies will be discussed.

7.1 FERROMAGNETIC RESONANCE

7.1.1 Equation of Motion

In Chapter 5 we stated the equation of motion of the magnetization with three different phenomenological forms for the damping terms. Repeated below, these equations are, in order of historical appearance, the Landau-Lifshitz (LL) equation, the Bloch-Bloembergen (BB) equation, and the Gilbert (G) equation:

LL:
$$\frac{d\mathbf{M}}{dt} = \gamma(\mathbf{M} \times \mathbf{H}) + \frac{\lambda}{M^2} \mathbf{M} \times (\mathbf{M} \times \mathbf{H}) \tag{7.1}$$

BB:
$$\frac{d\mathbf{M}}{dt} = \gamma(\mathbf{M} \times \mathbf{H}) - \frac{\hat{\mathbf{x}}M_x + \hat{\mathbf{y}}M_y}{\tau_2} - \hat{\mathbf{z}}\frac{(M_z - M)}{\tau_1} \tag{7.2}$$

G:
$$\frac{d\mathbf{M}}{dt} = \gamma(\mathbf{M} \times \mathbf{H}) + \frac{\alpha}{M} \mathbf{M} \times \frac{d\mathbf{M}}{dt} \tag{7.3}$$

Although the three damping terms are different in form, these equations to a first approximation are equivalent at low signal levels provided the damping is sufficiently small, i.e., $\alpha^2 \ll 1$. Because of historical precedence, the LL equation is the most widely used of the three. On the other hand the G equation is the most convenient one from a mathematical standpoint. That this is so can be easily demonstrated as follows: Let $\mathbf{H}$ and $\mathbf{M}$ be given by

$$\mathbf{H} = \hat{\mathbf{z}}H_z + \mathbf{h}e^{j\omega t} \tag{7.4}$$

$$\mathbf{M} = \hat{\mathbf{z}}M_0 + \mathbf{m}e^{j\omega t} \tag{7.5}$$

Substituting these expressions into Eq. (7.3), we obtain

$$j\omega\mathbf{m} = \alpha_e M_0(\hat{\mathbf{z}} \times \mathbf{h}) + (\omega_0 + j\omega\alpha)(\hat{\mathbf{z}} \times \mathbf{m}) \tag{7.6}$$

where $\omega_0 = -\gamma_e H_z$ is the Larmor precessional frequency of the electron in the static field H_z. Note that the damping factor α only appears in the combination $\omega_0 + j\omega\alpha$. Thus, in the initial and intermediate stages of our calculation, we can simplify matters by ignoring the damping term altogether provided we replace $-\gamma_e H_z = \omega_0$ by ω_0', which is defined as equal to $\omega_0 + j\omega\alpha$. Because of this mathematical simplicity, we shall use the Gilbert form when dealing with ferromagnetic resonance at low signal levels.

The microwave magnetic field and magnetization components $\mathbf{h}$ and $\mathbf{m}$ can be decomposed as

$$\mathbf{h} = \hat{\mathbf{x}}h_x + \hat{\mathbf{y}}h_y + \hat{\mathbf{z}}h_z \tag{7.7}$$

$$\mathbf{m} = \hat{\mathbf{x}}m_x + \hat{\mathbf{y}}m_y + \hat{\mathbf{z}}m_z \tag{7.8}$$

Substituting these equations into Eq. (7.6), we obtain the component equations:

$$j\omega m_x = -\omega_0'm_y - \gamma_e M_0 h_y \tag{7.9}$$

$$j\omega m_y = \gamma_e M_0 h_x + \omega_0'm_x \tag{7.10}$$

$$j\omega m_z = 0 \tag{7.11}$$

Solving for m_x and m_y as a function of h_x and h_y, we obtain

$$m_x = -\frac{\omega_0'\gamma_e M_0}{\omega_0'^2 - \omega^2} h_x - \frac{j\omega\gamma_e M_0}{\omega_0'^2 - \omega^2} h_y \tag{7.12}$$

$$m_y = \frac{j\omega\gamma_e M_0}{\omega_0'^2 - \omega^2} h_x - \frac{\omega_0'\gamma_e M_0}{\omega_0'^2 - \omega^2} h_y \tag{7.13}$$

Letting

$$\chi = \frac{\omega_0' \omega_M}{\omega_0'^2 - \omega^2} \tag{7.14}$$

$$\kappa = \frac{-\omega \omega_M}{\omega_0'^2 - \omega^2} \tag{7.15}$$

where
$$\omega_M = -\gamma_e M_0 \tag{7.16}$$

We can simplify Eqs. (7.11)–(7.13) to read

$$\begin{aligned}
m_x &= \chi h_x - j\kappa h_y \\
m_y &= j\kappa h_x + \chi h_y \\
m_z &= 0
\end{aligned} \tag{7.17}$$

Rewriting these equations in matrix form, we have

$$\begin{bmatrix} m_x \\ m_y \\ m_z \end{bmatrix} = \begin{bmatrix} \chi & -j\kappa & 0 \\ j\kappa & \chi & 0 \\ 0 & 0 & 0 \end{bmatrix} \begin{bmatrix} h_x \\ h_y \\ h_z \end{bmatrix} \tag{7.18}$$

From this matrix a susceptibility tensor $\|\chi\|$, defined by the relation $\mathbf{m} = \|\chi\|\mathbf{h}$, can be given:

$$\|\chi\| = \begin{bmatrix} \chi & -j\kappa & 0 \\ j\kappa & \chi & 0 \\ 0 & 0 & 0 \end{bmatrix} \tag{7.19}$$

where χ and κ are in general complex. Letting $\chi = \chi' - j\chi''$ and $\kappa = \kappa' - j\kappa''$, we readily find from Eqs. (7.14) and (7.15) the expressions for χ', χ'', κ', and κ'':

$$\begin{aligned}
\chi' &= \frac{\omega_M \omega_0 [\omega_0^2 - \omega^2(1 - \alpha^2)]}{[\omega_0^2 - \omega^2(1 + \alpha^2)]^2 + 4\omega^2 \omega_0^2 \alpha^2} \\[4pt]
\chi'' &= \frac{\omega_M \omega \alpha [\omega_0^2 + \omega^2(1 + \alpha^2)]}{[\omega_0^2 - \omega^2(1 + \alpha^2)]^2 + 4\omega^2 \omega_0^2 \alpha^2} \\[4pt]
\kappa' &= \frac{-\omega_M \omega [\omega_0^2 - \omega^2(1 + \alpha^2)]}{[\omega_0^2 - \omega^2(1 + \alpha^2)]^2 + 4\omega^2 \omega_0^2 \alpha^2} \\[4pt]
\kappa'' &= \frac{-2\omega_M \omega_0 \omega^2 \alpha}{[\omega_0^2 - \omega^2(1 + \alpha^2)]^2 + 4\omega^2 \omega_0^2 \alpha^2}
\end{aligned} \tag{7.20}$$

where we recall that $\omega_0' = \omega_0 + j\omega\alpha$. In this expression α^2 can usually be neglected compared to unity in the combinations $\omega^2(1 - \alpha^2)$ and $\omega^2(1 + \alpha^2)$ for all relevant ratios of ω_0/ω.

In our study of paramagnetic resonance the r.f. field at the sample was assumed to be circularly polarized, consistent with historical practice. Here, in the study of ferromagnetic resonance, it is customary to assume $\mathbf{h}$ to be in general elliptically polarized. In this case we have seen that the relationship between $\mathbf{m}$ and $\mathbf{h}$ is governed by a second-rank tensor. We expect, of course, that if $\mathbf{h}$ is circularly polarized, the equivalent scalar susceptibility should be given by expressions like those of Eqs. (6.25) and (6.26). That this is indeed the case can be further justified by noting the identity of Eqs. (6.3) and (7.6) for the case of circular polarization, provided τ_2 of the Bloch equations is set equal to $1/\omega\alpha$ of the Gilbert equation.

The equivalent scalar susceptibility for the general case of elliptical polarization can be easily found by diagonalizing the tensor in Eq. (7.19), i.e., by setting $\|\chi\|\mathbf{h}$ equal to $\lambda\mathbf{h}$ where λ is the eigenvalue or the equivalent susceptibility. Carrying out this calculation, we easily find the equivalent susceptibilities to be

$$\chi_+ = \chi - \kappa \tag{7.21}$$

$$\chi_- = \chi + \kappa \tag{7.22}$$

where $+$ and $-$ refer to positive (resonance) and negative polarization, respectively.[1] From Eqs. (7.14) and (7.15) and the relation $\omega_0' = \omega_0 + j\omega\alpha$, we easily find the following expressions:

$$\chi_+' = \frac{\omega_M(\omega_0 - \omega)}{(\omega_0 - \omega)^2 + (\omega\alpha)^2} \tag{7.23a}$$

$$\chi_+'' = \frac{\omega_M\omega\alpha}{(\omega_0 - \omega)^2 + (\omega\alpha)^2} \tag{7.23b}$$

Similarly, χ_-' and χ_-'' are given by

$$\chi_-' = \frac{\omega_M(\omega_0 + \omega)}{(\omega_0 + \omega)^2 + (\omega\alpha)^2} \tag{7.24a}$$

$$\chi_-'' = \frac{\omega_M\omega\alpha}{(\omega_0 + \omega)^2 + (\omega\alpha)^2} \tag{7.24b}$$

Comparing Eq. (7.23a) with (6.25) and (7.23b) with (6.26), we see that they are, respectively, identical provided that we set $\omega\alpha = 1/\tau_2$. For this reason the universal curves of Fig. 6.1 for the components of the susceptibility are also applicable to the ferromagnetic case here. Likewise, the linewidth ΔH, in analogy with Eq. (6.22), is given by

$$\Delta H = \frac{2\omega\alpha}{|\gamma_e|} \tag{7.25}$$

In a common experimental arrangement to be described later the sample is placed against the center of the back wall of a rectangular cavity and χ rather than $\chi + \kappa$ or $\chi - \kappa$ is measured. In this case the linewidth is given by the field separation between $\frac{1}{2}\chi_{max}''$ points where χ_{max}'' refers to the value of χ'' at resonance. A little mathematical manipulation involving Eq. (7.20) will show that as in the circularly polarized case and to the order of α^2, the resonance frequency ω_0 and linewidth are still given, respectively, by ω and $2\omega\alpha/|\gamma_e|$. This result is, of course, what one should expect since the resonance phenomenon involves one sense of circular polarization only; the presence of magnetic

---

[1]To show this, we need only find the eigenvectors associated with Eq. (7.18) in the usual way.

field of the opposite sense will cause relatively little absorption at and near resonance.

In the derivation of the susceptibility tensor in Eq. (7.19), we have implicitly assumed that H_z is sufficiently large so that the ferromagnetic material is saturated in the direction of $\hat{z}H_z$; i.e., the static component of **M** is equal to $\hat{z}M_0$ as assumed in Eq. (7.5). If H_z is not large enough, the sample will in general be divided into ferromagnetic domains within each of which the magnetic moments are aligned, as discussed in Section 4.6. In the absence of any applied field, the magnetization in these domains will be randomly oriented with respect to each other. When a small H_0 is applied, the boundaries of these domains will move in such a way as to expand the domains favorably oriented with respect to $\mathbf{H}_0$ and to contract those unfavorably oriented with respect to it. Thus, we can expect the component of magnetization along the $\mathbf{H}_z$ direction to be a function of H_z, up to a value $H_z = H_{sat}$, which is the value of H_z required to saturate the material. Furthermore, if an r.f. field is superimposed, we can expect the components of the susceptibility tensor to depend on the spatial average of the magnetization in the z direction.

For $\omega_0/\omega \ll 1$ and $\alpha^2 \ll 1$, Eq. (7.20) can be simplified to read

$$
\begin{aligned}
\chi' &= -\left(\frac{\omega_0}{\omega}\right)\frac{|\gamma|\langle M_0 \cos\theta\rangle}{\omega} \\
\chi'' &= \alpha\frac{|\gamma|\langle M_0 \cos\theta\rangle}{\omega} \\
\kappa' &= \frac{|\gamma|\langle M_0 \cos\theta\rangle}{\omega} \\
\kappa'' &= -2\alpha\left(\frac{\omega_0}{\omega}\right)\frac{|\gamma|\langle M_0 \cos\theta\rangle}{\omega}
\end{aligned}
\tag{7.26}
$$

where M_0 has been replaced by the spatial average $\langle M_0 \cos\theta\rangle$ and θ is the angle between the local magnetization and $\mathbf{H}_z$.[2] Note that $\langle M_0 \cos\theta\rangle$ vs H_z gives the ordinary magnetization or hysteresis curves. If we assume that Eq. (7.26) holds for $H_z < H_{sat}$, we can sketch the components of χ_+ and χ_- as in Fig. 7.1 (dotted); the solid portions of the curves were sketched using Eq. (7.23), which is applicable to cases where the sample is saturated.

The curves of Fig. 7.1 are strictly applicable only if $\omega/|\gamma| \gg H_{sat}$. If this is not the case, χ'_+, χ''_+, etc., will be given by the complete expressions (7.23) and (7.24) except that ω_M should be replaced by $|\gamma|\langle M_0 \cos\theta\rangle$. In general, if $\omega/|\gamma| < H_{sat}$ and $\Delta H \gtrsim \omega/|\gamma|$, we can expect the resonance line to be comparatively broader. Note that all components of the susceptibility approach zero as $H_z \to 0$. In other words the initial susceptibility is always zero. Actually, this is not always the case, especially at lower frequencies, since domain wall relaxation and domain rotation resonance can occur in the absence of H_0, giving

---

[2]This follows from the fact that $M_0 \cos\theta$ is the component of **M** along $\mathbf{H}_z$ for any given θ; to obtain the magnetization for the sample as a whole we must average $M_0 \cos\theta$ over all θ. See also the brief note by G. I. Rado, *Phys. Rev.* **89**, 529 (1953).

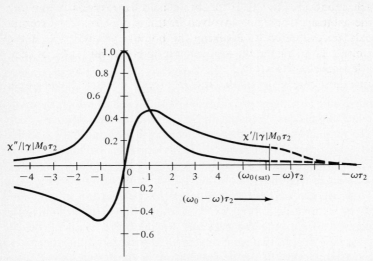

Figure 7.1 Normalized susceptibility components vs normalized frequency deviation from resonance. Solid portion: saturated state; dotted portion: unsaturated state.

rise to a finite initial susceptibility. Such details were not included in our previous discussion, in which we could account for the unsaturated state merely by replacing M_0 by $\langle M_0 \cos \theta \rangle$ and concerned ourselves not at all with the detail dynamics of domain magnetization and domain walls in r.f. fields. When domain wall relaxation and domain rotation resonance effects are important, they may give rise to appreciable losses in the unsaturated state. This phenomenon will be further discussed in Chapter 9, on ferrites.

7.1.2 Internal and External Fields

The expressions for χ', χ'', κ', and κ'' given by Eq. (7.20) were derived by implicitly concerning ourselves with magnetic fields inside the medium. Since external magnetic fields are more readily measured or ascertained, it is frequently more convenient to deal with external rather than internal fields. For ellipsoidal samples that are small compared to a wavelength, the relation between the internal and external fields can be written very simply:

$$
\begin{aligned}
h_x &= h_{xe} - N_x m_x \\
h_y &= h_{ye} - N_y m_y \\
H_z &= H_0 - N_z M_0
\end{aligned}
\tag{7.27}
$$

where we have used the demagnetizing field expression given by Eq. (4.67). h_x, h_y, and H_z are the internal field components while h_{xe}, h_{yes} and H_0 are the external field components. N_x, N_y, and N_z are, respectively, demagnetization factors in the x, y, and z directions. If the ellipsoidal sample is not small compared to a wavelength, Eq. (7.27) is no longer strictly applicable. In that case the effects of propagation in the sample must also be considered. This can be done by substituting the susceptibility tensor (7.19) based on internal fields into the Maxwell equations (2.1)–(2.4) to obtain field expressions within the sample. Then the arbitrary constants involved in the expressions for external and internal fields are evaluated by invoking the boundary conditions at the surface of the sample. This calculation is in general quite difficult and has been done for only a few special cases.[3,4]

Substituting the expressions for h_x and h_y given by Eq. (7.27) into Eq. (7.17), we have

$$m_x = \chi(h_{xe} - N_x m_x) - j\kappa(h_{ye} - N_y m_y)$$
$$m_y = \chi(h_{ye} - N_y m_y) + j\kappa(h_{xe} - N_x m_x)$$

(7.28)

Solving Eqs. (7.28) simultaneously for m_x and m_y, we find

$$m_x = \frac{\chi(1 + \chi N_y) - \kappa^2 N_y}{D} h_{xe} \frac{-j\kappa}{D} h_{ye}$$

$$m_y = \frac{j\kappa}{D} h_{xe} + \frac{\chi(1 + \chi N_x) - \kappa^2 N_x}{D} h_{ye}$$

(7.29)

where
$$D = (1 + \chi N_x)(1 + \chi N_y) - \kappa^2 N_x N_y$$

(7.30)

The resonance frequency of an ellipsoidal sample can be determined by setting D given by Eq. (7.30) equal to zero. Since χ and κ as well as the frequency ω are expected to be complex for a lossy medium, the calculation of the resonance frequency and Q from Eq. (7.30) will be a tedious one. However, for small damping ($\alpha^2 \ll 1$) or high $Q(\gg 1)$, the resonance frequency should deviate little from its value in the absence of damping, whether our calculation is based on internal or external fields. Setting $\alpha = 0$ in the expressions for χ and κ given by Eq. (7.20) and substituting the resulting simplified expression into Eq. (7.29), we find

$$m_x = \frac{\omega_M(\omega_0 + \omega_M N_y)}{D'} h_{xe} - \frac{j\omega\omega_M}{D'} h_{ye}$$

$$m_y = \frac{j\omega\omega_M}{D'} h_{xe} + \frac{\omega_M(\omega_0 + \omega_M N_x)}{D'} h_{ye}$$

(7.31)

where
$$D' = (\omega_0 + N_x\omega_M)(\omega_0 + N_y\omega_M) - \omega^2$$

(7.32)

Recalling that $\omega_0 = -\gamma H_z$ and $\omega_M = -\gamma M_0$, we find, on setting $D' = 0$, the following expression for the resonance frequency ω_r:

$$\omega_r = |\gamma|\sqrt{(H_z + N_x M_0)(H_z + N_y M_0)}$$

(7.33)

In terms of the applied static field H_0, this equation becomes

$$\omega_r = |\gamma|\sqrt{[H_0 + (N_x - N_z)M_0][H_0 + (N_y - N_z)M_0]}$$

(7.34)

At this point, it is relevant to recall that the local field acting on a given dipole moment in a cubic ferromagnet is given by the sum of three terms: the applied,

[3]R. F. Soohoo, *J. Appl. Phys.* **31**, 218S (1960).

[4]R. F. Soohoo and P. Christensen, *J. Appl. Phys.* **40**, 1565 (1969).

the demagnetizing, and the Lorentz fields. Whereas the applied and demagnetizing field enters explicitly into Eq. (7.34), the Lorentz field, like the exchange field, does not. As discussed in Section 4.6, the Lorentz and exchange fields are always parallel to **M** and act to align the magnetic moments in a sample together against thermal agitation. Were it not for the exchange and Lorentz fields, it would not be justified to represent the magnetization by **M**; thus, the Lorentz and exchange fields do enter into the resonance condition (7.34), but not explicitly. Looked at another way, the exchange and Lorentz fields, being parallel to **M**, can produce no torque on **M**. It thus follows from the Gilbert equation,

$$\frac{d\mathbf{M}}{dt} = \gamma(\mathbf{M} \times \mathbf{H}) + \frac{\alpha}{M}\, \mathbf{M} \times \frac{d\mathbf{M}}{dt} \tag{7.35}$$

that these fields can have no direct effect on the dynamic behavior of **M**.

On the other hand **M** need not be parallel to one of the axes of the ellipsoidal sample in the dynamic case. In the derivation of Eq. (7.34) we resolve **M** into three components parallel to the three axes of the ellipsoid. Each component of **M** will give rise to a pole distribution at the surface of the sample and a resultant demagnetizing field oppositely directed to this component of **M**. Similarly, **H** need not be parallel to **M** in the dynamic case either.

7.1.3 Magnetic Anisotropy

In the derivation of resonance condition (7.34) we implicitly neglected the effect of magnetic anisotropy. This is well justified for the case of a saturated polycrystalline sample since the anisotropy axes in that case are randomly oriented. For a single crystal or for a crystal with induced anisotropy, this is not the case. For example, in the case of magnetocrystalline anisotropy, the free energy density is given by the expression

$$E = K_0 + K_1(\alpha_1^2\alpha_2^2 + \alpha_2^2\alpha_3^2 + \alpha_3^2\alpha_1^2) + K_2\alpha_1^2\alpha_2^2\alpha_3^2 \tag{7.36}$$

For simplicity, consider only the first-order anisotropy energy density E_1. From Fig. 7.2 it can be shown that $\alpha_1 = \sin\theta\cos\phi$, $\alpha_2 = \sin\theta\sin\phi$, and $\alpha_3 = \cos\theta$. Thus, Eq. (7.36) reduces to

$$E_1 = K_1\sin^2\theta(\sin^2\theta\sin^2\phi\cos^2\phi + \cos^2\theta) \tag{7.37}$$

It is clear from this expression that if θ is held constant, E_1 will attain its minimum value $K_1\sin^2\theta\cos^2\theta$ at $\phi = 0, \pi/2, \pi, 3\pi/2$ along the $\pm x$ and $\pm y$ directions. Similarly, for constant ϕ, E_1 attains its minimum value 0 at $\theta = 0, \pi$ or along the $\pm z$ directions. Thus, if **M** is displaced from these equilibrium positions, we would expect a tendency for it to return to them. This restoring tendency can be represented by an equivalent anisotropy field oriented along one of the three orthogonal crystalline axes.

The equivalent anisotropy field $\mathbf{H}^e$ is defined such that the torque exerted on the sample by such a field is equal to that exerted because of the anisotropy

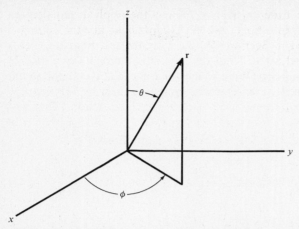

Figure 7.2 Coordinate system for determination of direction cosines. Note that x, y, and z correspond to the axes of a cubic crystal.

energy. It is often appropriate to take $\mathbf{H}^e$ as being parallel to the x and y axes of the sample. Furthermore, if the x and y components of $\mathbf{H}^e$ are expressed in terms of effective demagnetization factors N_x^e and N_y^e defined by the relations

$$H_x^e = -N_x^e M_x \tag{7.38}$$

and
$$H_y^e = -N_y^e M_y \tag{7.39}$$

Kittel's resonance equation (7.34) can now be modified to include the effect of anisotropy.[5] Thus,

$$\omega_r = |\gamma| \sqrt{[H_0 + (N_x + N_x^e - N_z)M_0][H_0 + (N_y + N_y^e - N_z)M_0]} \tag{7.40}$$

Let us now formulate the general and specific expressions for N_x^e and N_y^e under representative experimental situations. To begin with, the torque $\boldsymbol{\tau}$ due to a force $\mathbf{F}$ is given by the relation

$$\boldsymbol{\tau} = \mathbf{r} \times \mathbf{F} \tag{7.41}$$

If $\mathbf{F}$ is conservative, it can be expressed as minus the gradient of an energy E:

$$\mathbf{F} = -\nabla E \tag{7.42}$$

In spherical coordinates $\nabla E = \hat{\mathbf{r}}\, \partial E/\partial r + \hat{\boldsymbol{\theta}}(1/r)\, \partial E/\partial\theta + \hat{\boldsymbol{\phi}}(1/r \sin\theta)\, \partial E/\partial\phi$, and so Eqs. (7.41) and (7.42) can be combined to yield

$$\boldsymbol{\tau} = \hat{\mathbf{x}}\left(\sin\phi\, \frac{\partial E}{\partial\theta} + \cot\theta \cos\phi\, \frac{\partial E}{\partial\phi}\right)$$

$$+\, \hat{\mathbf{y}}\left(-\cos\phi\, \frac{\partial E}{\partial\theta} + \cot\theta \sin\phi\, \frac{\partial E}{\partial\phi}\right) - \hat{\mathbf{z}}\, \frac{\partial E}{\partial\phi} \tag{7.43}$$

[5]C. Kittel, *Phys. Rev.* **73**, 155(1948).

where we have resolved τ into its cartesian components. But τ can also be equated to $\mathbf{M} \times \mathbf{H}^e$ to yield

$$\tau = \hat{\mathbf{x}}(m_y H_z^e - M_z H_y^e) + \hat{\mathbf{y}}(M_z H_x^e - m_x H_z^e) + \hat{\mathbf{z}}(m_x H_y^e - m_y H_x^e) \tag{7.44}$$

Equating corresponding components of τ given by Eqs. (7.43) and (7.44) and noting that $m_x = M_0 \sin \theta \cos \phi$, $m_y = M_0 \sin \theta \sin \phi$, and $M_z = M_0 \cos \theta$, we finally find

$$M_0 \sin \theta \sin \phi \, H_z^e - M_0 \cos \theta \, H_y^e = \sin \phi \frac{\partial E}{\partial \theta} + \cot \theta \cos \phi \frac{\partial E}{\partial \phi}$$

$$M_0 \cos \theta \, H_x^e - M_0 \sin \theta \cos \phi \, H_z^e = -\cos \phi \frac{\partial E}{\partial \theta} + \cot \theta \sin \phi \frac{\partial E}{\partial \phi} \tag{7.45}$$

$$M_0 \sin \theta \cos \phi \, H_y^e - M_0 \sin \theta \sin \phi \, H_x^e = -\frac{\partial E}{\partial \phi}$$

With no loss of generality, we can set $H_z^e = 0$. Then we find

$$N_x^e = \frac{1}{M_0^2} \left[\frac{1}{\sin \theta \cos \theta} \frac{\partial E}{\partial \theta} - \frac{\tan \phi}{\sin^2 \theta} \frac{\partial E}{\partial \phi} \right]$$

$$N_y^e = \frac{1}{M_0^2} \left[\frac{1}{\sin \theta \cos \theta} \frac{\partial E}{\partial \theta} + \frac{\cot \phi}{\sin^2 \theta} \frac{\partial E}{\partial \phi} \right] \tag{7.46}$$

where Eqs. (7.38) and (7.39) were also used.[6] Once $E(\theta, \phi)$ is given for a particular problem, Eq. (7.46) can be used to find N_x^e and N_y^e. Inserting these expressions for N_x^e and N_y^e into Eq. (7.40), the influence of anisotropy on ω_r can be found.

7.2 LINEWIDTH CONTRIBUTIONS

As in paramagnetic resonance, two types of relaxation, spin-lattice and spin-spin, contribute to a finite ferromagnetic resonance linewidth. Again, as delineated in our discussion of paramagnetic resonance linewidth, sources of spin-lattice relaxation include (1) direct interaction between the spin system and the thermal electromagnetic radiation or photon radiation bath, (2) indirect interaction between the spin system and the phonon radiation bath, and (3) modulation of the ligand field by lattice vibrations. In addition, relaxation of the spin system may also occur because of its interaction with the conduction electrons in a ferromagnetic metal. The introduction of spin-lattice interaction, characterized by the spin-lattice relaxation time τ_1, limits the lifetime of an excited state of magnetization and thus gives rise to a finite linewidth.

At low signal levels spin-lattice interaction can occur due to the interaction between thermally generated spin waves (to be discussed below) and lattice

---

[6]An examination of Eq. (7.45) shows that if $\partial E/\partial \phi = 0$, it is permissible to set H_x^e and H_y^e, instead of H_z^e, equal to zero.

vibrations. The interaction in this case can be due to the tensor part of the dipole-dipole coupling, including the effect of lattice vibrations on the position vector $\mathbf{r}_{jk}$ between atoms j and k. Alternatively, the interaction can be attributed to the magnetoelastic energy. An interaction between spin waves and conduction electrons in ferromagnetic metals by spin-spin and spin-orbit effects can also exist. However, relaxation times calculated based on any of the above processes are too small to account for the resonance linewidth observed for ferromagnets at low signal levels.[7]

Spin-spin relaxation rather than spin-lattice relaxation appears to account for ferromagnetic resonance linewidth at low signal levels. The ordinary dipole-dipole coupling that gives rise to a range of effective values at different lattice sites and a spin flip phenomenon, discussed in connection with paramagnetic resonance, appears to be too weak to account for the observed anisotropy. A short-range (nearest-neighbor) pseudodipolar force that has the same structure as the tensor part of the ordinary dipole force with an interaction strength about a hundred times greater removes this difficulty. In addition, pseudo-quadrupole coupling can give rise to a line broadening by spin-spin relaxation. When the exchange energy is large compared to the Zeeman splitting and in the presence of dipolarlike interactions, the states of single wave number k will no longer be eigenstates of the system. The correct eigenstates are mixtures of the states of various k. The mixing effect, caused by the fluctuating dipolarlike fields, breaks down the selection rule $\Delta n_0 = \pm 1$, $\Delta n_k = 0$ ($k \neq 0$), and transition can take place over a group of states into which the $k = 0$ state has been admixed.

Because of the presence of strong exchange forces in a ferromagnet that tend to hold neighboring spins parallel, line broadening due to dipolelike forces can be greatly reduced. Under favorable conditions the linewidth can be less than an oersted.

In a ferromagnetic metal the microwave skin depth can be quite small. Within this skin depth the microwave magnetization is no longer uniform. Hence the exchange coupling between neighboring spins can lead to spin-spin relaxation and thus play a role in determining the linewidth.

In general, the experimental ferromagnetic resonance line is neither truly Lorentzian nor truly Gaussian, but that corresponding to $\|\chi\|$ given by Eq. (7.19) is strictly Lorentzian.

7.3 SPIN WAVES

Saturation effects in ferromagnets are found to occur at power levels much lower (100–200 times lower) than that predicted from a linear theory. It is not possible, as with dilute paramagnets, to drive a ferromagnetic system so hard

[7]E. Abrahams, "Relaxation Processes in Ferromagnetism," *Advances in Electronics and Electron Physics*, Vol. 6, Academic, New York, 1954, p. 47.

that the magnetization component M_z can be reduced to zero or reversed. The ferromagnetic resonance excitation breaks down into spin wave modes before the magnetization vector can be rotated by more than a few degrees from its initial direction.

If a ferromagnetic ellipsoid is placed in uniform r.f. and static magnetic fields[8] at 0°K and if boundary effects are neglected, uniform precession may be excited whereby the direction of spins throughout the sample volume does not change from point to point.[9] However, at finite temperatures or in nonuniform fields, or if the physical dimensions of the samples are such that boundary effects cannot be neglected, the spin direction may have a spatial dependence. Macroscopically, we may expand the r.f. magnetization in the form of a Fourier series of plane waves with each spatial harmonic identified as a "spin wave."

If the r.f. magnetization **m** varies from point to point continuously and with continuous first derivatives, we can expand **m** at frequency ω as

$$\mathbf{m}(\mathbf{r}, t) = \sum_{\mathbf{k}} \mathbf{m}_k e^{j(\omega t - \mathbf{k} \cdot \mathbf{r})} \tag{7.47}$$

where **k** is the wave vector of wave number k and the summation is over all possible modes **k**. If Eq. (7.47) is solved in conjuction with the equation of motion of the magnetization and Maxwell's equations with appropriate boundary conditions, we can, in principle at least, evaluate the unknown coefficients $\mathbf{m}_k$ for any experimental situation and sample configuration.

According to Eq. (4.53), the exchange energy E_{ex} of an assembly of spins is given by

$$E_{ex} = -2 \sum_{i,j}' J_{i,j} \mathbf{S}_i \cdot \mathbf{S}_j \tag{7.48}$$

where J_{ij} is the exchange integral between spins $\mathbf{S}_i$ and $\mathbf{S}_j$. $\sum_{i,j}'$ indicates that the summation is over nearest neighbors only and with the prime indicating the exclusion of self-energy terms $i = j$. Equation (7.48) is a semiclassical expression in that $\mathbf{S}_i$ and $\mathbf{S}_j$ are considered as vectors rather than as operators. In any macroscopic ferromagnetic sample, the number of spins N is very very large so that the corresponding number of allowable net spin orientations $2NS + 1$ is also extremely large. It is then not unreasonable to treat Ss as classical vectors (rather than as operators) whose directions are not quantized.

Rewriting Eq. (7.48), we have

$$E_{ex} = -2JS^2 \sum_{j} \cos \theta_{ij} \tag{7.49}$$

with θ_{ij} being the angle between spins i and j. Here we have assumed that only the exchange interactions between spin i and its nearest neighbors j are important. Furthermore, we have assumed that $J_{ij} = J$ is independent of j and that $|\mathbf{S}| = S$.

---

[8]Strictly speaking, the r.f. field would be nearly uniform only if the sample dimensions are much smaller than a wavelength in the medium.

[9]Zero-point oscillation is being ignored here.

Alternatively, Eq. (7.49) can be expressed in terms of the direction cosines α:

$$\cos \theta_{ij} = \hat{\mathbf{u}}_i \cdot \hat{\mathbf{u}}_j = \alpha_{1i} \alpha_{1j} + \alpha_{2i} \alpha_{2j} + \alpha_{3i} \alpha_{3j} \tag{7.50}$$

where $\hat{\mathbf{u}}_i$ and $\hat{\mathbf{u}}_j$ are unit vectors in the directions of spins i and j. α_{1i}, α_{2i}, and α_{3i} are equal, respectively, to the cosine of the angle (direction cosines) of $\hat{\mathbf{u}}_i$ relative to the cartesian x, y, and z axes. If the angle between the $\hat{\mathbf{u}}_i$ and $\hat{\mathbf{u}}_j$ is small, the direction cosine of $\hat{\mathbf{u}}_j$ may be expanded in a Taylor series in the direction cosine of $\hat{\mathbf{u}}_i$:

$$\alpha_{1i} \alpha_{1j} = \alpha_{1i} \left[\alpha_{1i} + \mathbf{r}_{ij} \cdot \nabla \alpha_{1i} + \tfrac{1}{2}(\mathbf{r}_{ij} \cdot \nabla)^2 \alpha_{1i} + \cdots \right] \tag{7.51}$$

where $\mathbf{r}_{ij} = \hat{\mathbf{i}} x_{ij} + \hat{\mathbf{j}} y_{ij} + \hat{\mathbf{k}} z_{ij}$ is the displacement vector directed from spin i to spin j.

To obtain an expression for E_{ex} given by Eq. (7.48) in terms of the αs, we need to sum expression (7.51) over the nearest neighbors j. For cubic crystals, terms like $\sum_j \mathbf{r}_{ij} \cdot \nabla \alpha_{1i}$ and cross terms from $\sum_j \tfrac{1}{2}(\mathbf{r}_{ij} \cdot \nabla)^2 \alpha_{1j}$ like $\sum_j x_{ij} y_{ij} (\partial^2 \alpha_{1j} / \partial x_{ij} \partial y_{ij})$ are zero by virtue of symmetry. Thus,

$$\sum_j \cos \theta_{ij} = z + \tfrac{1}{2}\alpha_{1i} \frac{\partial^2 \alpha_{1i}}{\partial x_{ij}^2} \sum_j x_{ij}^2 + \tfrac{1}{2}\alpha_{1i} \frac{\partial^2 \alpha_{1i}}{\partial y_{ij}^2} \sum_j y_{ij}^2 + \tfrac{1}{2}\alpha_{1i} \frac{\partial^2 \alpha_{1i}}{\partial z_{ij}^2} \sum_j z_{ij}^2$$

$$+ \tfrac{1}{2}\alpha_{2i} \frac{\partial^2 \alpha_{2i}}{\partial x_{ij}^2} \sum_j x_{ij}^2 + \cdots \tag{7.52}$$

Since for cubic crystals,

$$\sum_j x_{ij}^2 = \sum_j y_{ij}^2 = \sum_j z_{ij}^2 = \frac{1}{3} \sum_j r_{ij}^2$$

Eq. (7.52) simplifies to

$$\sum_j \cos \theta_{ij} = z + \frac{1}{6} \sum_j r_{ij}^2 \, \hat{\mathbf{u}} \cdot \nabla^2 \hat{\mathbf{u}} \tag{7.53}$$

where subscript i on $\hat{\mathbf{u}}_i$ has been omitted. Substituting this expression into Eq. (7.48) and retaining only the variable part of the energy,

$$E_{ex} = -\frac{JS^2}{3} \sum_j r_{ij}^2 \, \hat{\mathbf{u}} \cdot \nabla^2 \hat{\mathbf{u}} \tag{7.54}$$

For sc, bcc, or fcc crystals, $\sum_j r_{ij}^2 = 6a^2$ where a is the lattice spacing. Thus, the exchange energy for one cube of edge a is

$$E_{ex} = -2JS^2 a^2 \, \hat{\mathbf{u}} \cdot \nabla^2 \hat{\mathbf{u}} \tag{7.55}$$

Accordingly, since there are $1/a^3$ cubes per unit volume, the exchange energy per unit volume is

$$E_{ex} = -\frac{2JS^2}{a} \, \hat{\mathbf{u}} \cdot \nabla^2 \hat{\mathbf{u}}$$

$$= -\frac{A}{M^2} \mathbf{M} \cdot \nabla^2 \mathbf{M} \tag{7.56}$$

where $A = 2JS^2/a$ is known as the exchange constant. This energy may also be written as $-\frac{1}{2}\mathbf{M} \cdot \mathbf{H}_{ex}$ in which $\mathbf{H}_{ex}$ is the equivalent exchange field acting on $\mathbf{M}$ and the factor of $\frac{1}{2}$ enters because E_{ex} represents a self-energy. Comparing Eq. (7.56) with this expression, it is seen that

$$\mathbf{H}_{ex} = \frac{2A}{M^2}\nabla^2\mathbf{M} \tag{7.57}$$

If this expression for $\mathbf{H}_{ex}$ is introduced into the Gilbert equation of motion (7.3), we have

$$\dot{\mathbf{M}} = \gamma\mathbf{M} \times \left(\mathbf{H} + \frac{\alpha}{M}\dot{\mathbf{M}} + \frac{2A}{M^2}\nabla^2\mathbf{M}\right) \tag{7.58}$$

This expression must be solved in conjunction with Maxwell's equation:

$$\nabla \times \mathbf{h} = \epsilon\frac{\partial e}{\partial t} + \sigma\mathbf{e} \tag{7.59}$$

$$\nabla \times \mathbf{e} = -\mu_0\frac{\partial\mathbf{h}}{\partial t} - \mu_0\frac{\partial\mathbf{m}}{\partial t} \tag{7.60}$$

where we have also used the relations $\mathbf{i} = \sigma\mathbf{e}$, $\mathbf{d} = \epsilon\mathbf{e}$, and $\mathbf{b} = \mu_0(\mathbf{h} + \mathbf{m})$. Taking the curl of Eq. (7.59) and solving the resulting equation for $\nabla \times \mathbf{e}$ for substitution into Eq. (7.60), we find

$$\nabla^2\mathbf{h} - \mu_0\left(\epsilon\frac{\partial}{\partial t} + \sigma\right)\frac{\partial\mathbf{h}}{\partial t} = \mu_0\left(\epsilon\frac{\partial}{\partial t} + \sigma\right)\frac{\partial\mathbf{m}}{\partial t} - \nabla(\nabla \cdot \mathbf{m}) \tag{7.61}$$

where we have used the relation $\nabla \cdot \mathbf{b} = \nabla \cdot [\mu_0(\mathbf{h} + \mathbf{m})] = 0$. Note that if $\mathbf{m} = 0$, Eq. (7.61) reduces to the well-known wave equation for a dielectric medium of dielectric constant ϵ and conductivity σ. If $\mathbf{m}$ is of the form given by Eq. (7.47), $\mathbf{h}$ should likewise be given by a similar expression:

$$\mathbf{h} = \sum_{\mathbf{k}} \mathbf{h}_k e^{j(\omega t - \mathbf{k} \cdot \mathbf{r})} \tag{7.62}$$

Substituting Eqs. (7.47) and (7.62) into Eq. (7.61), we have

$$\mathbf{h}_k = \frac{\omega^2\mu_0\epsilon_e\mathbf{m}_k - \mathbf{k}(\mathbf{k} \cdot \mathbf{m})}{k^2 - \omega^2\mu_0\epsilon_e} \tag{7.63}$$

where $\epsilon_e = \epsilon(1 + \sigma/j\omega)$ is the equivalent dielectric constant. Combining Eqs. (7.58) and (7.63), we obtain two linear algebraic equations in the x and y component of $\mathbf{m}$.[10] Setting the determinant of these equations equal to zero, we obtain a secular equation biquadratic in k^2. Thus, in general, the magnetization $\mathbf{m}$ can be expressed in terms of a linear combination of eight plane spin waves. The expression for $\mathbf{h}$ can then in turn be obtained from Eq. (7.63). Imposing the ex-

[10]R. F. Soohoo, *Phys. Rev.* 120, 1978 (1960); *Magnetic Thin Films*, Harper & Row, New York, 1965, p. 208.

change boundary condition (to be discussed below) and the usual requirement of the continuity of the tangential components of **h** and **e** at the surfaces of the sample, we can find the surface impedance Z_s of the specimen. Plotting Z_s vs H_0 will give an absorption spectrum that can be compared with experiment.

7.4 LINE SPECTRUM OF THIN FILMS

The mathematics involved in the calculation of the absorption spectrum is quite complicated. For this reason it is more instructive for us to restrict our attention to the calculation of the line spectrum, in which case α and σ are both assumed zero. In this case the equation of motion, (7.58), becomes

$$\dot{\mathbf{M}} = \gamma \mathbf{M} \times \left(\mathbf{H} + \frac{2A}{M^2} \nabla^2 M \right) \tag{7.64}$$

If we further assume that the spin wave number **k** is z directed or perpendicular to the surfaces of a film [see Fig. 7.3(a)], the expression for **H** and **M** becomes

$$\mathbf{H} = \hat{\mathbf{z}}(H_0 - M)$$
$$\mathbf{M} = \hat{\mathbf{z}}M + \mathbf{m}e^{j(\omega t - kz)} \tag{7.65}$$

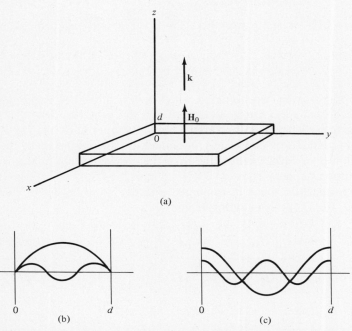

Figure 7.3 (a) z-directed spin wave in a thin film. Exchange boundary conditions for (b) pinned and (c) unpinned cases.

In the expression for **H** we have neglected the time-varying component given by Eq. (7.63). This is equivalent to the neglect of propagation effects, an assumption justified for spin waves propagating through a thin film. Substituting Eq. (7.65) in Eq. (7.64), we obtain

$$j\omega m_x - \gamma \left(H_0 - M + \frac{2A}{M} h^2 \right) m_y = 0$$

$$\gamma \left(H_0 - M + \frac{2A}{M} k^2 \right) m_x + j\omega m_y = 0$$

(7.66)

Setting the determinant of these equations equal to zero, we find

$$\omega = |\gamma| \left(H_0 - M + \frac{2A}{M} k^2 \right)$$

(7.67)

The line spectrum can be found from Eq. (7.67) provided the values of k are known. Two possible cases present themselves, one for which $\mathbf{m} = 0$ and another for which $\partial \mathbf{m}/\partial z = 0$ at the surface of the film. These respective assumptions are known as the pinned and unpinned exchange boundary conditions. Referring to Fig. 7.3(b), we see that for the pinned case

$$kd = n\pi \qquad n = 1, 3, 5, \ldots$$

(7.68)

while for the unpinned case [see Fig. 7.3(c)]

$$kd = n\pi \qquad n = 0, 2, 4, 6, \ldots$$

(7.69)

Substituting Eqs. (7.68) or (7.69) into Eq. (7.67), we obtain the dispersion relation

$$\omega = |\gamma| \left[H_0 - M + \frac{2A}{M} \left(\frac{n\pi}{d} \right)^2 \right]$$

(7.70)

where n is either odd or even depending on whether $\mathbf{m}$ is pinned or unpinned at the surface. Accordingly the spectra at a fixed ω is given by Fig. 7.4.

It is of course possible that the surface conditions are such that $\mathbf{m}$ is neither completely pinned nor completely unpinned. In this general case the line will lie at a lower or higher field value than the corresponding modes of the pinned and unpinned cases, respectively.

The exchange constant A can be evaluated with the aid of Eq. (7.70). For modes i and j, (see Fig. 7.4) this equation reads

$$\omega = |\gamma| \left[H_{0i} - M + \frac{2A}{M} \left(\frac{\pi}{d} \right)^2 n_i^2 \right]$$

(7.71)

$$\omega = |\gamma| \left[H_{0j} - M + \frac{2A}{M} \left(\frac{\pi}{d} \right)^2 n_j^2 \right]$$

(7.72)

Subtracting Eq. (7.72) from Eq. (7.71), we have

$$H_{0i} - H_{0j} = \frac{2A}{M} \left(\frac{\pi}{d} \right)^2 (n_j^2 - n_i^2)$$

(7.73)

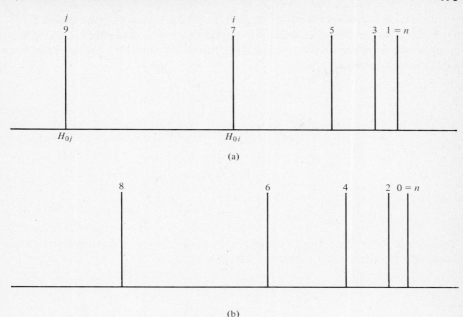

Figure 7.4 Line resonance spectra for thin films.

Thus, by measurement of the resonance field separation $H_{0i} - H_{0j}$ between spin wave modes i and j, A can be found provided n_j and n_i can be identified.

Because of its high conductivity, a metallic magnetic film has rather limited microwave applications owing to its high eddy current loss. For this reason Permalloy ($\sim 80\%$ Ni, 20% Fe) films (100–10,000 Å thick) are presently used mainly for computer applications, although their intrinsic properties, such as magnetization, g value, and linewidth (related to its switching time as a computer memory), can be measured in ferromagnetic resonance experiments at microwave frequencies.

For nonconducting films made of ferrites or garnets, the eddy current loss at microwave frequencies is relatively small. Thus, they may be used at microwave frequencies to build nonreciprocal devices, for example. However, there has been little exploitation of this principle.

7.5 ANTIFERROMAGNETIC RESONANCE

For simplicity, let us consider a uniaxial antiferromagnet with the magnetization on the two sublattices given by $\mathbf{M}_1$ and $\mathbf{M}_2$ directed in the $+z$ and $-z$ directions, respectively. In this case the anisotropy energy density is given by $E_k(\theta_1) = K_1 \sin^2 \theta_1$ where θ_1 is the angle between M_1 and the z axis (a similar expression exists for $\mathbf{M}_2$). It can be shown from equation of motion (7.3) that as far as resonance in low signals is concerned, the anisotropy effect can be represented by an effective field $H_k = 2K_1/M$ along the z axis. Noting further that the exchange or molecular field for sublattices 1 and 2 are given, respectively, by $-\lambda \mathbf{M}_2$ and $-\lambda \mathbf{M}_1$ (the minus sign appears because $\mathbf{M}_1$ and $\mathbf{M}_2$ are oppositely directed in an antiferromagnet), equation of motion (7.3) in its linear approximation becomes

$$\frac{dm_{1x}}{dt} = \gamma[m_{1y}(\lambda M + H_k) - M(-\lambda m_{2y})]$$

$$\frac{dm_{1y}}{dt} = \gamma[M(-\lambda m_{2x}) - m_{1x}(\lambda M + H_k)] \tag{7.74}$$

$$\frac{dm_{2x}}{dt} = \gamma[m_{2y}(-\lambda M - H_k) - (-M)(-\lambda m_{1y})]$$

$$\frac{dm_{2y}}{dt} = \gamma[(-M)(-\lambda m_{1x}) - m_{2x}(-\lambda M - H_k)] \tag{7.75}$$

Letting the time dependence be of the form $e^{j\omega t}$, we can define the positive circularly polarized components of $\mathbf{M}_1$ and $\mathbf{M}_2$ as

$$m_1^+ = m_{1x} - jm_{1y}$$
$$m_2^+ = m_{2x} - jm_{2y} \tag{7.76}$$

Substituting Eq. (7.76) into Eqs. (7.74) and (7.75), we find

$$j\omega m_1^+ = j\gamma[m_1^+(H_k + \lambda M) + m_2^+(\lambda M)]$$
$$j\omega m_2^+ = j\gamma[m_2^+(H_k + \lambda M) + m_1^+(\lambda M)] \tag{7.77}$$

Setting the determinant of these equations equal to zero, we find the antiferromagnetic resonance frequency to be

$$\omega^2 = \gamma^2 H_k(H_a + 2H_{ex}) \tag{7.78}$$

where $H_{ex} = \lambda M$. In general $H_{ex} \gg H_a$. Hence, Eq. (7.78) simplifies to

$$\omega = \gamma\sqrt{2H_k H_{ex}} \tag{7.79}$$

For MnF_2, an antiferromagnet, $H_k \simeq 8.8$ kOe and $H_{ex} \simeq 540$ kOe at $0°K$. It follows that the effective field $\sqrt{2H_k H_{ex}}$ is about 100 kOe and the corresponding response frequency is 280 GHz, a very high value indeed. Because antiferromagnets can self-resonate at very high microwave frequencies, they have been used to build millimeter wave isolators.

PROBLEMS

7.1 If for a field applied in the x direction, say, there are components of magnetization not only in the x direction but in the y and z directions as well, then the susceptibility is a tensor rather than a scalar. What attribute in a magnetic material gives rise to this phenomenon?

7.2 (a) Show that the eigenvalues of the matrix in Eq. (7.18) are χ_+ and χ_- given by Eqs. (7.21) and (7.22).

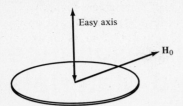

Easy axis

H_0

Figure 7.5 Resonance in a bubble film.

(b) Show that χ_+ and χ_- are, respectively, the equivalent susceptibilities of the positive and negative circularly polarized waves.

7.3 Find the value of ω_0 at which χ''_{ext} given by $-\mathrm{Im}(m_x/h_{xe})$ of Eq. (7.29) attains a maximum, and then show that for $\alpha^2 \ll 1$ the value of ω_r is given by Eq. (7.34).

7.4 Sketch to scale the dotted portions of the χ'_+ and χ''_+ vs $(\omega - \omega_0)/\omega\alpha$ curves of Fig. 7.1 if the magnetization for $0 \leq H_z \leq H_{sat}$ can be approximated by the expression $M_0(1 - e^{-H_z/5H_{sat}})$.

7.5 Sketch the χ'_+ and χ''_+ vs $(\omega - \omega_0)/\omega\alpha$ curves for the case in which the medium is unsaturated. Specifically, let $\omega/|\gamma| = H_{sat}/2$ and assume that $M_z = M_0(1 - e^{-H_z/5H_{sat}})$ as in Problem 7.4.

7.6 Although the r.f. field h is much less than the applied field H_0, the demagnetizing fields due to $\mathbf{h}\,(= -\hat{x}N_xM_0 - \hat{y}N_yM_0)$ are of the same order as that due to H_0 $(= -\hat{z}N_zM_0)$ [see Eq. (7.34)]. Explain why this is reasonable.

7.7 (a) Figure 7.5 shows a magnetic bubble film with its uniaxial anisotropy easy axis oriented perpendicular to the film. If the anisotropy constant is K, find the resonance frequency for the case of $\mathbf{H}_0$ applied perpendicular to the film.

   (b) Similarly, find the resonance frequency for the case of $\mathbf{H}_0$ applied in the plane of the film.

Chapter 8

Ferrimagnetic Resonance
and Materials

Much of the material contained in Chapter 7 will be applicable to the case of ferrimagnetic resonance since the equations of motion [Eqs. (7.1)–(7.3)] do not explicitly include eddy current losses due to finite conductivity. In this chapter we shall therefore utilize such results as the susceptibility tensor given by Eq. (7.19) as the starting point of our study of microwave propagation through an anisotropic medium. Subsequently, various nonreciprocal ferrite and garnet devices and their related applications will be discussed.

8.1 MICROWAVE PROPAGATION THROUGH INFINITE ANISOTOPIC MEDIA

Consider a wave of wave number $\hat{\mathbf{k}}\gamma$, making an angle θ with the applied field or z axis, propagating through an infinite anisotopic medium (see Fig. 8.1). In this case the fields have the dependence $e^{j\omega t - \gamma \hat{\mathbf{k}} \cdot \mathbf{r}}$. Substituting this expression into the Maxwell equations (2.1)–(2.4), we have

$$\frac{\gamma}{j\omega} \mathbf{E} \times \hat{\mathbf{k}} = -\mathbf{b} \tag{8.1}$$

$$\frac{\gamma}{j\omega} \mathbf{h} \times \hat{\mathbf{k}} = \mathbf{D} \tag{8.2}$$

Combining Eq. (8.2) with Eq. (2.9), we have

$$\mathbf{E} = \frac{1}{\epsilon} \left(\frac{\gamma}{j\omega} \mathbf{h} \times \hat{\mathbf{k}} \right) \tag{8.3}$$

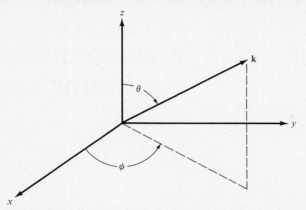

Figure 8.1 Coordinate system for wave propagation through an anisotropic medium.

Substituting Eq. (8.3) into Eq. (8.1), we find

$$b = \frac{1}{\epsilon}\left(\frac{\gamma}{j\omega}\right)^2 [\mathbf{h} - (\mathbf{h} \cdot \hat{\mathbf{k}})\hat{\mathbf{k}}]$$

(8.4)

where the vector identity $(\mathbf{h} \times \hat{\mathbf{k}}) \times \hat{\mathbf{k}} = (\mathbf{h} \cdot \hat{\mathbf{k}})\hat{\mathbf{k}} - \mathbf{h}$ has been used for convenience. But, according to Eq. (7.19), $\mathbf{b}$ is also given by

$$\mathbf{b} = \mu_0 \begin{bmatrix} \mu & -j\kappa & 0 \\ j\kappa & \mu & 0 \\ 0 & 0 & 1 \end{bmatrix} \begin{pmatrix} h_x \\ h_y \\ h_z \end{pmatrix}$$

(8.5)

where the components of the permeability tensor $\mu = 1 + \chi$ and κ are defined by Eq. (7.20). Equating the right-hand sides of Eqs. (8.4) and (8.5), we find the following component equations:

$$\mu_0(\mu h_x - j\kappa h_y) = \frac{1}{\epsilon}\left(\frac{\gamma}{j\omega}\right)^2 [h_x - (\sin\theta\cos\phi\, h_x + \sin\theta\sin\phi\, h_y$$
$$+ \cos\theta\, h_z)\sin\theta\cos\phi]$$

$$\mu_0(j\kappa h_x + \mu h_y) = \frac{1}{\epsilon}\left(\frac{\gamma}{j\omega}\right)^2 [h_y - (\sin\theta\cos\phi\, h_x + \sin\theta\sin\phi\, h_y$$
$$+ \cos\theta\, h_z)\sin\theta\sin\phi]$$

(8.6)

$$\mu_0 h_z = \frac{1}{\epsilon}\left(\frac{\gamma}{j\omega}\right)^2 [h_z - (\sin\theta\cos\phi\, h_x + \sin\theta\sin\phi\, h_v$$
$$+ \cos\theta\, h_z)\cos\theta]$$

Setting the determinant of these equations equal to zero, we find the eigenvalues for γ:

$$\gamma_\pm = j\omega(\mu_0\epsilon)^{1/2}$$
$$\times \left\{\frac{(\mu^2 - \mu - \kappa^2)\sin^2\theta + 2\mu \mp [(\mu^2 - \mu - \kappa^2)^2\sin^4\theta + 4\kappa^2\cos^2\theta]^{1/2}}{2[(\mu - 1)\sin^2\theta + 1]}\right\}^{1/2}$$

(8.7)

Note that $\gamma_\pm$ is independent of the angle ϕ. It can be shown that the solutions corresponding to γ_+ and γ_- represent two elliptically polarized waves propagating in the z direction, but with different velocities.

8.2 LONGITUDINAL-FIELD CASE

If **k** is directed along the $\mathbf{H}_0$ axis ($\theta = 0$), expression (8.7) for $\gamma_\pm$ reduces to

$$\gamma_\pm = j\omega[\mu_0(\mu \mp \kappa)\epsilon]^{1/2} \tag{8.8}$$

The equivalent relative permeabilities $\mu - \kappa$ and $\mu + \kappa$ are sketched in Fig. 8.2. In this sketch Eq. (7.26) for the unsaturated regions as well as Eqs. (7.23) and (7.24) for the saturated regions are used. Specifically, in the unsaturated region, where $\omega_0 \ll \omega$ and for $\alpha^2 \ll 1$, components of $\mu - \kappa$ and $\mu + \kappa$ are given by

$$\mu'_+ = 1 - \left(1 + \frac{\omega_0}{\omega}\right)\frac{|\gamma|\langle M_0 \cos\theta\rangle}{\omega} \tag{8.9}$$

$$\mu''_+ = \alpha\left(1 + 2\frac{\omega_0}{\omega}\right)\frac{|\gamma|\langle M_0 \cos\theta\rangle}{\omega} \tag{8.10}$$

and
$$\mu'_- = 1 + \left(1 - \frac{\omega_0}{\omega}\right)\frac{|\gamma|\langle M_0 \cos\theta\rangle}{\omega} \tag{8.11}$$

$$\mu''_- = \alpha\left(1 - 2\frac{\omega_0}{\omega}\right)\frac{|\gamma|\langle M_0 \cos\theta\rangle}{\omega} \tag{8.12}$$

where Eq. (7.26) has been used. Similarly, for the saturated region, we have

$$\mu'_+ = 1 + \frac{\omega_M(\omega_0^2 - \omega^2)(\omega_0 + \omega)}{(\omega_0^2 - \omega^2)^2 + 4\omega^2\omega_0^2\alpha^2} \tag{8.13}$$

$$\mu''_+ = \frac{\omega_M\omega\alpha(\omega_0 + \omega)^2}{(\omega_0^2 - \omega^2)^2 + 4\omega^2\omega_0^2\alpha^2} \tag{8.14}$$

and
$$\mu'_- = 1 + \frac{\omega_M(\omega_0^2 - \omega^2)(\omega_0 - \omega)}{(\omega_0^2 - \omega^2)^2 + 4\omega^2\omega_0^2\alpha^2} \tag{8.15}$$

$$\mu''_- = \frac{\omega_M\omega\alpha(\omega_0 - \omega)^2}{(\omega_0^2 - \omega^2)^2 + 4\omega^2\omega_0^2\alpha^2} \tag{8.16}$$

Except in the region of resonance ($\omega \simeq \omega_0$), the terms involving α^2 in Eqs. (8.13)–(8.16) can be neglected. In this case

$$\mu'_+ = 1 + \frac{|\gamma|M_0}{|\gamma|H_0 - \omega} \tag{8.17}$$

$$\mu''_+ = \frac{|\gamma|M_0\omega\alpha}{(\gamma H_0 - \omega)^2} \tag{8.18}$$

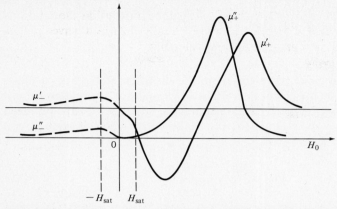

Figure 8.2 Equivalent permeability components vs the static field for propagation parallel to the static field.

and
$$\mu'_- = 1 + \frac{|\gamma| M_0}{|\gamma| H_0 + \omega} \tag{8.19}$$

$$\mu''_- = \frac{|\gamma| M_0 \omega \alpha}{(|\gamma| H_0 + \omega)^2} \tag{8.20}$$

since not only $\alpha^2 \ll 1$, but also α itself is frequently small compared to 1. In this case the behavior of devices far away from resonance will be controlled by Eqs. (8.17) and (8.19) for μ'_+ and μ'_-.

8.3 FARADAY ROTATION

According to Eq. (8.8), the propagation constant is different for waves of opposite circular polarization. Since in general elliptically polarized waves can be decomposed into positive and negative circularly polarized waves of unequal amplitude, let us consider the combination of two electric fields of the form

$$e_{t+} = E_{t+} \, e^{j\theta - \gamma_+ z} \, e^{j\omega t} \tag{8.21}$$

$$e_{t-} = E_{t-} \, e^{-j\theta - \gamma_- z} \, e^{j\omega t} \tag{8.22}$$

Because of the anisotropic nature of the medium, γ_+ and γ_- are in general unequal. For small applied H_0, however, γ''_+ and γ''_- are sufficiently small in value (see Fig. 8.2), provided low field losses can be neglected, that we can assume $\gamma_+ = j\beta_+$ and $\gamma_- = j\beta_-$ where β_+ and β_- are, respectively, the phase constants of the positive and negative circularly polarized waves. Thus, the superposition of these two waves gives the electric field for a TE mode:

$$e_t = E_t (e^{j\theta - j\beta_+ z} + e^{-j\theta - j\beta_- z}) e^{j\omega t} \tag{8.23}$$

where we have assumed $E_{t+} = E_{t-} = E_t$ for simplicity. At $z = 0$

$$e_t(0) = 2E_t \cos \theta \, e^{j\omega t} \tag{8.24}$$

We see that $e_t(0)$ is maximum for $\theta = 0$, π. For definiteness, we shall let $\theta = 0$ determine the direction of polarization at $z = 0$. For $z = 1$

$$e_t(z) = 2E_t[e^{-j(1/2)(\beta_+ + \beta_-)z}]\cos[\theta + \tfrac{1}{2}(\beta_- - \beta_+)z]e^{j\omega t} \qquad (8.25)$$

As expected, the combined wave travels in the direction of $\mathbf{H}_0$ with a phase velocity determined by the algebraic average of β_+ and β_-. More interestingly, the polarization corresponding to the maximum amplitude of $e_t(z)$ and obtained by setting $\theta + \tfrac{1}{2}(\beta_- - \beta_+)z$ equal to zero is seen to have been rotated; the Faraday rotation for length z is

$$\theta(z) = \left(\frac{\beta_+ - \beta_-}{2}\right)z \qquad (8.26)$$

Combining Eqs. (8.8), (8.9), and (8.11) with Eq. (8.26), we find for the unsaturated region where $\omega_0 \ll \omega$,

$$\theta(z) = \frac{\omega\sqrt{\mu_0\epsilon}}{2}\left\{\left[1 - \left(1 + \frac{\omega_0}{\omega}\right)\frac{|\gamma|\langle M_0\cos\theta\rangle}{\omega}\right]^{1/2} \right. \\ \left. - \left[1 + \left(1 - \frac{\omega_0}{\omega}\right)\frac{|\gamma|\langle M_0\cos\theta\rangle}{\omega}\right]^{1/2}\right\} \qquad (8.27)$$

If the second term in each bracket is much smaller than unity, Eq. (8.27) can be simplified by means of the binomial series to obtain

$$\theta(z) \simeq -\frac{\sqrt{\mu_0\epsilon}}{2}\,|\gamma|\langle M_0\cos\theta\rangle z \qquad (8.28)$$

Thus, for a given z, θ is directly proportional to $\langle M_0\cos\theta\rangle$. On the other hand a plot of $\langle M_0\cos\theta\rangle$ vs H_0 has just the shape of a hysteresis loop. Accordingly, θ vs H_0 for a given z is as shown in Fig. 8.3.

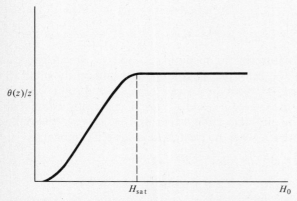

Figure 8.3   Faraday rotation per unit length vs the applied field.

8.4 TRANSVERSE-FIELD CASE

If $\mathbf{H}_0$ is transverse to the direction of propagation, we set $\theta = \pi/2$ in Eq. (8.7) to obtain

$$\gamma_\perp = j\omega(\mu_0\epsilon)^{1/2} \left(\frac{\mu^2 - \kappa^2}{\mu}\right)^{1/2} \tag{8.29}$$

$$\gamma_\parallel = j\omega(\mu_0\epsilon)^{1/2} \tag{8.30}$$

The field dependence of $\mu_\perp = (\mu^2 - \kappa^2)/\mu$ is given in Fig. 8.4.

Here it is more appropriate to designate γ_+ by $\gamma_\perp$ and γ_- by $\gamma_\parallel$ since these γs correspond to the propagation constants for the field components perpendicular and parallel to the applied field. That the normal modes are as described can be shown by substituting

$$\mathbf{h} = (\hat{\mathbf{x}}h_x e^{-\gamma_\perp z} + \hat{\mathbf{y}}h_y e^{-\gamma_\perp z} + \hat{\mathbf{z}}h_z e^{-\gamma_\parallel z})e^{j\omega t} \tag{8.31}$$

into Eq. (8.6.) In addition to verifying that the expressions $\gamma_\perp$ and $\gamma_\parallel$ are as given by Eqs. (8.29) and (8.30) by this substitution, we find the relation between h_x and h_y to be

$$h_y = -j\frac{\kappa}{\mu} h_x \tag{8.32}$$

Substituting this relation into Eq. (7.17) and noting that $\mathbf{b} = \mu_0(\mathbf{h} + \mathbf{m})$, we have

$$\begin{aligned} b_x &= \mu_0(\mu h_x - j\kappa h_y) = \mu_0 \frac{\mu^2 - \kappa^2}{\mu} h_x \\ b_y &= \mu_0(j\kappa h_x + \mu h_y) = 0 \\ b_z &= \mu_0 h_z \end{aligned} \tag{8.33}$$

Whereas $\mathbf{h}$ is elliptically polarized, $\mathbf{b}$ is clearly linearly polarized in the x-y plane.

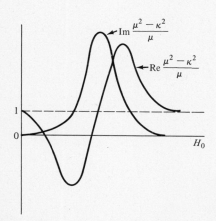

Figure 8.4 Equivalent permeability components vs the static field for propagation perpendicular to the static field.

8.5 RESONANCE ABSORPTION AND NONRECIPROCAL PHASE SHIFT

According to Eqs. (8.29) and (8.30) whereas $\gamma_{\parallel}$ is field-independent, γ is a function of H_0. The r.f. magnetic field component perpendicular to the static magnet field $\hat{z}H_0$, being elliptically polarized, can be decomposed into two counter-rotating circularly polarized components of unequal amplitude. The component that rotates in the same direction as the electron spin precession will give rise to resonance absorption while the component that rotates in the opposite direction does not. Thus, to maximize nonreciprocal propagation effect, it is necessary to locate the ferrite sample in a region of circular polarization in a waveguide.

In an empty rectangular waveguide of width a the magnetic field components for the TE_{10} fundamental mode are

$$h_x = jB\left(\frac{a}{\pi}\right)\beta \sin\frac{\pi x}{a}$$

$$h_y = -B\cos\frac{\pi x}{a}$$

(8.34)

These expressions follow from Eqs. (3.145) and (3.146) and the coordinate system depicted by Fig. 8.5; the direction of $\hat{z}$ is here chosen to correspond to the direction of an applied field when a ferrite sample is inserted. We note that h_x and h_y are in space and time quadrature, and that if, in addition, their magnitudes are equal, circular polarization will be assured. Equating $|h_x|$ and $|h_y|$ and recalling that $\beta^2 = \omega^2\mu_0\epsilon_0 - k_c^2$ where $k_c = \omega_c\sqrt{\mu_0\epsilon_0}$, we find the location x where the r.f. magnetic field is circularly polarized:

$$x_1 = \frac{a}{\pi}\tan^{-1}\frac{1}{\sqrt{1-(f_c/f)^2}}$$

(8.35)

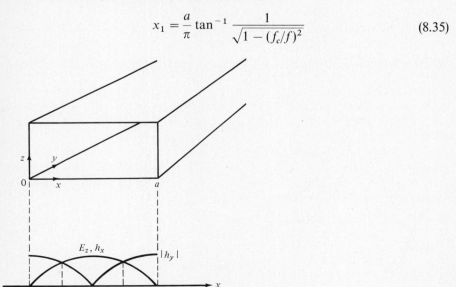

Figure 8.5   Field distribution of TE_{10} mode in a rectangular guide.

where $f_c = c/2a$ is the cutoff frequency of the waveguide. If a small ferrite slab is placed at x_1 or at $a - x_1$, nonreciprocal propagation can be expected.

The above analysis, based on empty waveguide fields, is not entirely correct, however. When a ferrite sample is inserted inside a waveguide, the field inside and in the vicinity of the ferrite sample will be significantly perturbed. In particular, the internal field in the sample, rather than the empty waveguide field at the location of the ferrite, must be circularly polarized to produce maximum nonreciprocal propagation. In practice, this means that the optimum location of the sample is not exactly given by Eq. (8.35). This and other aspects of the subject will be discussed in the next chapter, on ferrite and garnet devices.

8.6 FERRITES AND GARNETS[1]

Ferrite devices are used in virtually every microwave system and measurement. A resonance isolator prevents frequency pulling of microwave sources such as reflex Klystrons by virtue of its nonreciprocal attenuation while a circulator can perform switching functions by virtue of its nonreciprocal phase shift. When narrow-band or tunable devices are desired, garnet samples of small linewidth are frequently used. The crystal structures of ferrites and garnets have been studied in Section 4.8. In the present chapter their methods of preparation will be discussed. This will be followed in Chapter 9 by an examination of the characteristics of microwave ferrite devices and their application to actual systems.

Ferrites are ceramic ferromagnetic materials with the general chemical composition $MO \cdot Fe_2O_3$ where M is a divalent metal such as iron, manganese, magnesium, nickel, zinc, cadmium, cobalt, copper, or a combination of these. Although iron ferrite or magnetite ($FeO \cdot Fe_2O_3$ or Fe_3O_4) was known to the ancients as lodestone, the first usable modern ferrites were made only in recent times.[2] This development can be attributed in part to the demand for magnetic materials with low-core losses in carrier and in part to radio, television, and computer memory applications. Activities in the field of microwave ferrites began at a still later date. Polder first derived the permeability tensor in 1949, which laid the groundwork for the understanding of ferrite behavior at microwave frequencies.[3] In 1952 Hogan successfully constructed the first workable ferrite microwave gyrator.[4] Since then there has been a great deal of research and engineering activities in the microwave ferrite area.

Briefly stated, ferrites are made by sintering a mixture of various metallic oxides. During the sintering process the mixture crystallizes into a cubic structure like that of the mineral spinel $MgAl_2O_4$. The resulting ferrite has a hard, black, and nonporous ceramic appearance.

[1]For more details see R. F. Soohoo, *Theory and Application of Ferrites*, Prentice-Hall, Englewood Cliffs, NJ 1960; *IEEE Trans. Magnetics* **Mag-4**, 118 (1968); both with extensive references.

[2]J. L. Snoek, *Physica* **3**, 463 (1936); *Philips Tech. Rev.* **8**, 353 (1946); *New Developments in Ferromagnetic Materials*, Elsevier, New York, 1947.

[3]D. Polder, *Philos. Mag.* **40**, 99 (1949).

[4]C. L. Hogan, *Bell Syst. Tech. J.* **31**, 1 (1952).

A diagram of the unit cell of the spinel crystal structure is shown in Fig. 4.16, where the Mg^{2+}, Al^{3+}, and O_4^{2-} ions are represented by spheres of different sizes and shades. Note that the solid lines indicate the fourfold (tetrahedral) and sixfold (octahedral) coordination of the respective metallic ion positions. The metallic ion having four nearest oxygen neighbors is said to be at the A site while the one having six nearest oxygen neighbors is said to be at the B site. For zinc and cadmium ferrites the divalent metallic Zn^{2+} or Cd^{2+} ion is at the A site, while the two trivalent ferric ions $2Fe^{3+}$ are at the B site. This structure is the same as the normal spinel structure of Fig. 4.16. Most of the simple ferrites, e.g., $NiFe_2O_4$, however, are of the inverse spinel structure, in which one trivalent ferric ion Fe^{3+} is at the A site while the remaining trivalent ferric ion Fe^{3+} and the divalent metallic ion M^{2+} are at the B site.

There are notable exceptions to the $MO \cdot Fe_2O_3$ formula for ferrites. For example, Ferroxdure, a ferrite used for permanent magnets, has the chemical formula $BaO \cdot 6Fe_2O_3$. It has a hexagonal structure rather than the cubic crystal structure of the spinel. With proper orientation of the constituent single crystals, a high polycrystalline uniaxial anisotropy can be obtained. Because of its high-anisotropy field, it self-resonates with no externally applied field at about 50 GHz. Consequently, this material can be utilized to build light and compact resonance isolators at these high frequencies. Rare-earth garnets are types of ferrites with the general chemical formula $5Fe_2O_3 \cdot 3M_2O_3$, where M represents yttrium or some other rare-earth ions from samarium to lutecium. This structure differs from the spinel lattice of the conventional ferrites in several respects. Two of the chief structural differences are that the garnet has three types of lattice sites available to metallic ions, as compared to the two types of sites in the spinel, and that all of the possible sites in the garnet lattice are filled, as opposed to the half-empty B and the one-eighth-filled A sites of the spinel lattice. A two-dimensional view of a garnet unit cell is shown in Fig. 4.17. Because of their small linewidths, which give rise to a low threshold power for the excitation of spin waves, garnets can be used advantageously in nonlinear devices such as power limiters. A number of other ferrites or garnets have also been prepared, chiefly by doping or substitution.

8.7 FERRITE AND GARNET PREPARATION

The preparation of polycrystalline ferrites can best be illustrated by the flow-chart of Fig. 8.6.[5,6] It is seen that ferrites are made by sintering a mixture of various metallic oxides. Usually the exact composition, firing temperature, firing time, and furnace atmosphere required to produce ferrites with a particular set of characteristics are determined by repeated experimentation.

As shown in Fig. 8.6, 50 mol % of MO is mixed with 50 mol % of Fe_2O_3 in a wet mix using water and Aerosal or kerosene and Vasol Amine 220. The

[5]J. E. Pippin and C. L. Hogan, "The Preparation of Polycrystalline Ferrites for Microwave Applications," Gordon McKay Laboratory, Sci. Rep. 8, Harvard University, Cambridge, MA.

[6]D. L. Fresh, *Proc. IRE* **44**, 1303 (1956).

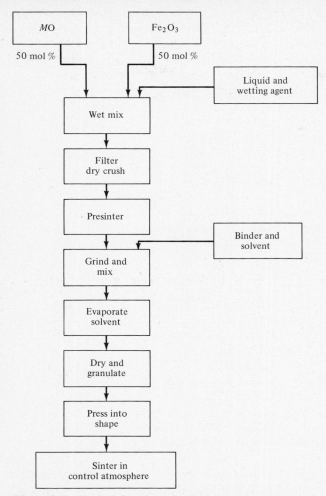

Figure 8.6 Typical flowchart for ferrite preparation.

wet mixing can be done in a ball mill or in a blender for several hours. Then the mix is filtered through funnels fitted with qualitative filter paper. The filtered cake is dried in air or in an oven heated to 110°C and then crushed through a 20- to 100-mesh screen. Next, the mixture is presintered in a furnace at 900°– 1100°C for 3–15 h. After the presintering operation, the mixture changes in color from red to dark gray or black, characteristic of the finished ferrites. The presintered powder is now ground and mixed with paraffin wax and water. Drying is done in air or in oven heated to 110°C. The granulating is accomplished by crushing the mix with a mortar and pestle and then forcing it through a screen. Then the granulated powder is dry pressed in a die with pressure ranging upward from 6000 psi. The last step in the ferrite preparation process is to place the ferrite in an alundum boat in a furnace and sinter it at 1000°–1400°C for a number of hours in an air or oxygen atmosphere.

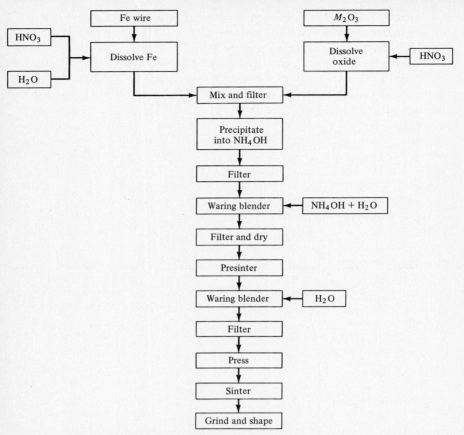

Figure 8.7   Flowchart for polycrystalline garnet preparation.

A coprecipitation process has been developed for the preparation of high-density polycrystalline garnets $5Fe_2O_3 \cdot 3M_2O_3$, where again M is an yttrium or a rare-earth ion ranging from samarium to lutecium.[7] The coprecipitation method of preparation has several advantages over the ceramic process for ferrite preparation that has been described. Extremely intimate mixing is achieved with a minimum of time and effort by combining the constituents in a liquid rather than a solid state, and with suitable precautions this atomic mixing is preserved on precipitation. This is particulary important in the preparation of mixed garnets, when small percentages of additional ions must be uniformly dispersed throughout the solid. High chemical purity and accurate control of stoichiometry are also facilitated by this process.

A flowchart of garnet preparation is shown in Fig. 8.7. Starting with the yttrium or rare earth in oxide form and the iron in metallic form, each of the constituent materials is first dissolved in nitric acid. The nitrate solutions are

[7]W. P. Wolf and G. P. Rodrigue, *J. Appl. Phys.* **29**, 105 (1958).

then mixed and filtered. Filtering at this stage removes any impurities of the starting materials, which are insoluble in nitric acid. In order to obtain narrow linewidth materials, it is necessary to use yttrium oxide of high purity, 99.9 + %. The hydroxides of iron and yttrium or rare earths are next formed by dropping the filtered nitrate solution into an excess of ammonium hydroxide. The solution is then filtered and the precipitate collected. The precipitate, together with the filter paper, is next finely divided by blending for 1–3 h in a dilute solution of ammonium hydroxide. After it is filtered and dried, the cake is presintered at temperatures ranging from 700° to 1000°C for about 8 h. The garnet compound, bright green in color, begins to form at presintering temperatures between 600° and 700°C. After the ball is milled or blended with water, the garnet compound is next filtered and sludge pressed into a desired shape with pressures ranging from 6000 to 40,000 psi. In the last step of the process the samples are fired in an oxygen atmosphere at temperatures ranging from 1300° to 1500°C for 6–15 h. Garnets should be fired on a platinum surface as they exhibit a tendency to react with most ceramics such as alundum boats. Bulk garnets sintered above 1250°C are black in appearance, very similar to ferrites.

Because of their narrow linewidths (tens of oersteds, as compared to hundreds of oersteds for ferrites) and relatively low magnetization (hundreds of gauss, as compared to thousands of gauss for ferrites), polycrystalline garnets virtually removed the low-frequency limit on microwave ferrite devices due to unsaturated low field losses below resonance. Because of their small power threshold for spin-wave excitation, single-crystal garnets, with linewidths as small as a fraction of an oersted, can be advantageously used in power limiters. In this connection we may also note that the relative dielectric constants of microwave ferrites and garnets are in the range of 8–16 while their loss tangents vary from 10^{-4} to 10^{-2}. The dielectric loss represents the limiting loss of nonresonant microwave ferrite devices such as the latching phase shifter.

Single crystals of ferrites or garnets are useful for the study of fundamental properties such as magnetic anisotropy. However, it is difficult to grow them large enough for practical applications in most ferrite devices. For details of their preparation, the reader is referred to the literature.[8–12]

8.8 FREQUENCY LIMITATIONS

Problems arise if we try to extend frequency coverage to the high and low microwave ranges. At high microwave frequencies very high magnetic fields are required for resonance, while at low microwave frequencies low field losses may limit device performance.

[8]J. W. Nielsen *J. Appl. Phys.* **31**, 51s (1960).

[9]L. I. Abernethy, T. H. Ramsey, Jr., and J. W. Ross, *J. Appl. Phys.* **32**, 376s (1967).

[10]A. Tanber, R. O. Savage, R. J. Gambino, and C. G. Whinfrey, *J. Appl. Phys.* **33**, 1381 (1962).

[11]W. Kunnmann, A. Wold, and E. Banks, *J. Appl. Phys.* **33**, 1364 (1962).

[12]E. D. Kolb, D. L. Wood, E. G. Spencer, and R. A. Laudise, *J. Appl. Phys.* **38**, 1027 (1967).

The high-frequency problem can be alleviated by the use of ferrites having a high-anisotropy field. For example, magnetic materials with hexagonal crystal structure whose preferred direction of magnetization is along the c axis are uniaxial with a uniaxial anisotropy field of up to 30,000 Oe. With the use of such ferrites, such as Ferroxdure or $Ni_2W + Zn_2W$, isolators can be built without any external field up to 110 GHz in the millimeter wave band. From about 2 mm to the submillimeter band, ferrites become impractical. Antiferromagnetic materials such as Cr_2O_3, with an internal field of about 50,000 Oe, may be then used to build an isolator at 150 GHz.

At low microwave frequencies, i.e., from 100 to 1000 MHz, both low field and resonance losses in the forward direction limit the reverse-to-forward attenuation ratio of an isolator. Low field losses exist if the field required to magnetize the ferrite totally (the saturation field) is comparable to the resonance field, a probable condition at low frequencies. If the ferrite linewidth is sufficiently large, resonance loss can also occur in the forward direction, since the resonance field at these low frequencies is very small. One solution to this problem is to use garnets that have a very low linewidth.

For high-average-power applications, a ferrite with a high Curie temperature, the temperature at which magnetization goes to zero, is required. For high-peak-power applications, a ferrite with large linewidth is desirable. Otherwise, spin waves may be excited to cause anomalous absorption behavior.

PROBLEMS

8.1 Derive Eq. (8.7).

8.2 Show that γ_+ and γ_- of Eq. (8.8) represent the equivalent permeabilities of circularly polarized waves.

8.3 Find the resonance frequency of the permeability $(\mu^2 - \kappa^2)/\mu$ depicted in Fig. 8.4.

8.4 Derive Eq. (8.32).

8.5 It turns out that the resonance frequency of $(\mu^2 - \kappa^2)/\mu$ of Problem 8.3 is not equal to the Larmor frequency $|\gamma|H_0$. Explain why this is possible.

Chapter 9

Ferrite and Garnet Devices

Microwave ferrite and garnet devices are numerous, but they can be broadly classified into five groups:

1. isolators,
2. circulators,
3. phase shifters,
4. switches, and
5. bandpass filters.

In this chapter representative devices in each of these categories will be discussed to illustrate the basic principles involved.

9.1 ISOLATORS

Before the appearance of the resonance isolator, the Faraday rotation effect was utilized to obtain isolation at microwave frequencies. This device, known as the Uniline,[1] has since been replaced by the resonance isolator because the former is more difficult to make and has a low power-carrying capacity. The structure of a resonance isolator is relatively simple; it can consist of one to four strips of magnetized ferrite placed at planes of circular polarization in a rectangular waveguide as shown in Fig. 9.1.

---

[1]R. F. Soohoo, *Theory and Application of Ferrites*, Prentice-Hall, Englewood Cliffs, NJ, 1960, p. 155.

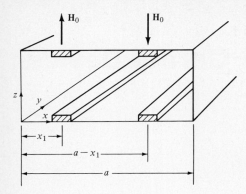

Figure 9.1 Resonance isolator ferrite and bias field configuration.

According to Eq. (8.35), the planes of circular polarization in an empty guide are given by

$$x_1 = \frac{a}{\pi} \tan^{-1} \frac{1}{\sqrt{(f/f_c)^2 - 1}} \tag{9.1}$$

Furthermore, an examination of Eq. (8.34) from which Eq. (8.35) was derived shows that h is negative circularly polarized at x_1; i.e., **h** rotates in the clockwise direction when viewed along the $+z$ direction, and positive circularly polarized at $a - x_1$. For this reason, if nonreciprocal propagation is desired, the dc fields at x_1 and $a - x_1$ must be along the z axis but oppositely directed. It is left as an exercise for the reader to show that the dc field orientations shown in Fig. 9.1 give low loss for energy propagated in the positive y direction and high loss for energy propagated in the negative y direction at ferrimagnetic resonance.

In Fig. 9.2 x_1/a are plotted as a function of normalized frequency f/f_c. Many optimum experimental locations of ferrites in typical resonance isolators are also shown for comparison with experiment. These data cover devices of many waveguide bands in a frequency range from 1 to 35 GHz. It is seen that the optimum ferrite location is, in general, about 4% closer to the guidewall than that predicted by theory, optimum ferrite location being defined as the location at which, if the ferrite samples were placed there, the maximum reverse-to-forward attenuation ratio would be found. Note that in actual ferrite isolators from one to four slabs of ferrite are used even though four slabs are shown in Fig. 9.1.

The discrepancy between theoretical and experimental x_1/a values, although small, is significant from an academic standpoint. When ferrite samples are inserted into the waveguide, the field distribution will obviously be disturbed and will no longer be given by the empty waveguide expressions [Eqs. (3.145)–(3.147)]. To obtain the exact field distribution, it is necessary to solve the boundary value problem indicated by the configuration of Fig. 9.1. This is a hopelessly complicated task: An infinity of modes is required to satisfy the boundary conditions because the ferrite samples do not extend across the entire guide. From an engineering standpoint an alternative approximate solution would be more useful. In this method the field distribution in the vicinity of

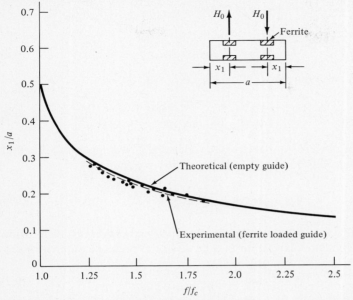

Figure 9.2 Point of circular polarization of a rectangular waveguide magnetic field vs frequency. Dots represent experimental points for several waveguide bands ranging from 1 to 35 GHz. Actual isolators use from one to four slabs of ferrite (four are shown here).

the ferrite is assumed to maintain its empty guide value. Because of the magnetic poles induced at the sample surfaces by the dc and r.f. fields, the fields inside the ferrite are nevertheless different from the corresponding values at the air-ferrite interfaces. For ellipsoidal samples that are small compared to the wavelength, the relation between external and internal fields is given by Eq. (7.27)[2]:

$$h_x = h_{xe} - N_x m_x \tag{9.2a}$$

$$h_y = h_{ye} - N_y m_y \tag{9.2b}$$

$$H_z = H_0 - N_z M_0 \tag{9.2c}$$

where h_{xe}, h_{ye} are, respectively, x and y r.f. field components at the ferrite-air interface while H_0 is the dc applied field. Correspondingly, h_x, h_y, H_z are the field components inside the ferrite. Likewise m_x, m_y, M_0 are the x, y, z components of the magnetization.[3] Finally, N_x, N_y, N_z are the demagnetization factors[4] along the x, y, z axes of the ellipsoid with $N_x + N_y + N_z = 1$.

[2]Since the sample is assumed ellipsoidal and small compared to a wavelength, h_x, h_y, h_z, m_x, m_y, M_0 are all independent of location inside the sample.

[3]See footnote 2.

[4]Values of N_x, N_y, N_z for ellipsoids of various shapes have been calculated by J. A. Osborn, *Phys. Rev.* **67**, 351 (1945). For nonellipsoids no N_x, N_y, N_z can strictly be defined as the demagnetizing fields would no longer be uniform. However, locally $N_x + N_y + N_z = 1$ even for these cases [see E. Schlomann, *J. Appl. Phys.* **33**, 2825 (1962)].

Combining Eqs. (9.2) and (7.17), we find

$$h_x = h_{xe} - N_x(\chi h_x - j\kappa h_y) \tag{9.3}$$

$$h_y = h_{ye} - N_y(j\kappa h_x + \chi h_y) \tag{9.4}$$

Solving Eqs. (9.3) and (9.4) simultaneously, we obtain

$$h_{xe} = \frac{(1 + N_x\chi)h_x - jN_x\kappa h_y}{(1 + N_y\chi)h_y - jN_y\kappa h_x} h_{ye} \tag{9.5}$$

If the internal field were to be circularly polarized, we must require that $h_x = \pm jh_y$. Inserting this condition into Eq. (9.5), we obtain

$$h_{xe} = \pm j\frac{1 + N_x(\chi \mp \kappa)}{1 + N_y(\chi \mp \kappa)} h_{ye} \tag{9.6}$$

Combining Eqs. (8.35) and (9.6), we finally obtain an expression for the location of the planes of circular polarization in a guide in the presence of ferrite slabs:

$$x_1 = \frac{a}{\pi} \tan^{-1}\left[\frac{r}{\sqrt{(f/f_c)^2 - 1}}\right] \tag{9.7}$$

where $r = [1 + N_x(\chi \mp \kappa)]/[1 + N_y(\chi \mp \kappa)]$. Note that Eq. (9.7) is, in general, different from the corresponding Eq. (9.1) for an empty waveguide.

In the derivation of Eq. (9.7) single-mode propagation was assumed. This is strictly justified only if the ferrite is assumed to be lossless. Otherwise coupling to other modes, propagating or otherwise, due to losses is unavoidable. Furthermore, the sample must be vanishingly thin and extending from top to bottom of the guide. If the sample is not vanishingly thin, the field inside the guide, especially near the ferrite, will be disturbed when the ferrite sample is inserted whereas the derivation of Eq. (9.7) implicitly assumed that no field disturbance occurs because of the presence of the ferrite. In any event we may recall that the concept of demagnetization factors is valid only if the sample is much smaller than a wavelength. Additionally, if the x dimension of the ferrite is not independent of z, an infinity of modes is required to satisfy the boundary conditions even in the absence of losses.

To consider further the applicability of Eq. (9.7) to actual situations, we note that $\chi - \kappa$ and $\chi + \kappa$, related to forward and reverse propagation, respectively, are in general unequal. Thus, the optimum ferrite locations for forward and reverse propagation in general differ, except for the case where the sample has cylindrical symmetry about the z axis ($N_x = N_y$ so that $r = 1$ for either case). However, if $N_x \simeq N_y \simeq 0$, roughly corresponding to many actual cases where the ferrite slabs of Fig. 9.1 are very thin compared to their width and length, then x_1 attains its empty-waveguide value. However, since N_x is always larger than N_y, r is nearly equal to but larger than unity. Equation (9.7) then says that x with the ferrite sample should be larger than without, in contradiction to the experimental results depicted in Fig. 9.2. To this we can only say that it is a result of the neglect of such factors as losses, field disturbance,

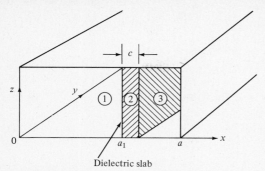

Figure 9.3 Dielectric-loaded waveguide.

finite sample size, sample dimension not z independent, etc., in the derivation of Eq. (9.7).

9.2 DIELECTRIC LOADING

Dielectric loading can significantly affect the performance of a resonance isolator; it can increase the reverse-to-forward absorption ratio as well as the bandwidth of an isolator. For this reason the technique of dielectric loading is probably the most important innovation in the isolator field since its inception. In this section we shall study the theory of dielectric loading and improvement in isolator performance.

To begin our discussion, we observe from Eq. (9.1) or (9.7) that the planes of circular polarization are zero in thickness. Thus, for a sample of finite x dimension, only part of it can be at resonance at a given frequency. This means that the isolation ratio would be smaller than the case where the field is circularly polarized throughout the sample, at least at a single frequency. Indeed, dielectric loading serves to make the fields inside the sample more circularly polarized over a broader frequency range and as a result can increase both the isolation ratio and bandwidth of isolators.

To determine the ellipticity, h_x/h_y, in the vicinity of the dielectric, let us consider the dielectric-loaded waveguide of Fig. 9.3. From the general field expressions (3.11), (3.30), and (3.31), we find the field components in regions 1–3 as follows[5]:

$$E_{z1} = A_1 \frac{\gamma_{0a} Z_{0a}}{k_{ca}} \sin k_{ca}x \tag{9.8}$$

$$h_{x1} = -A_1 \frac{\gamma}{k_{ca}} \sin k_{ca}x \tag{9.9}$$

$$h_{y1} = A_1 \cos k_{ca}x \tag{9.10}$$

[5]Note that for these expressions, the direction of propagation is assumed to be y rather than z.

and
$$E_{z2} = A_2 \frac{\gamma_{0d} Z_{0d}}{k_{cd}} \sin k_{cd}x + B_2 \frac{\gamma_{0d} Z_{0d}}{k_{cd}} \cos k_{cd}x \tag{9.11}$$

$$h_{x2} = -A_2 \frac{\gamma}{k_{cd}} \sin k_{cd}x - B_2 \frac{\gamma}{k_{cd}} \cos k_{cd}x \tag{9.12}$$

$$h_{y2} = A_2 \cos k_{cd}x - B \sin k_{cd}x \tag{9.13}$$

and
$$E_{z3} = A_3 \frac{\gamma_{0a} Z_{0a}}{k_{ca}} \sin k_{ca}(a - x) \tag{9.14}$$

$$h_{x3} = -A \frac{\gamma}{k_{ca}} \sin k_{ca}(a - x) \tag{9.15}$$

$$h_{y3} = -A_3 \cos k_{ca}(a - x) \tag{9.16}$$

where we have made use of the boundary conditions $E_{z1} = 0$ at $x = 0$ and $E_{z3} = 0$ at $x = a$. Also, in these expressions

$$k_{ca}^2 = \gamma_{0a}^2 + \gamma^2 = -\omega^2 \mu_0 \epsilon_0 + \gamma^2 \tag{9.17}$$
$$k_{cd}^2 = \gamma_{0d}^2 + \gamma^2 = -\omega^2 \mu_0 \epsilon + \gamma^2 \tag{9.18}$$

Imposing the boundary conditions

$$E_{z1} = E_{z2} \tag{9.19}$$
$$h_{y1} = h_{y2} \tag{9.20}$$

at $x = a_1$ and

$$E_{z2} = E_{z3} \tag{9.21}$$
$$h_{y2} = h_{y3} \tag{9.22}$$

at $x = a_1 + c$, we obtain five algebraic equations in the constants A_1, A_2, B_2, and A_3. Setting the determinant of these equations equal to zero, a characteristic equation from which γ can be determined is obtained. Once γ is determined, the field components h_x, h_y and the ellipticity ratio $e = |h_x|/|h_y|$ can be found.

For a centrally located slab, i.e., for $a_1 = \frac{1}{2}(a - c)$, $|h_x|$ and $|h_y|$ for unit power flow are plotted vs x/a in Fig. 9.4 for $\epsilon = 9\epsilon_0$ and several values of c/a. Note that the ellipticity e at the air-dielectric interfaces, where h_y changes abruptly, approaches closer and closer to unity as c/a increases. Indeed, for $c/a = 0.15$, the ellipticity is near unity for much of the air-filled space. This implies that an isolator can be built with thin ferrite slabs of width nearly equal to $\frac{1}{2}(a - c)$ and placed against the top and bottom guidewalls. This configuration has the advantage of close contact between the ferrite and metallic wall, which facilitates heat dissipation and results in a high power-carrying capacity. Loosely speaking, we can say that the planes of circular polarization in an empty guide are smeared into regions of near circular polarization when the guide is loaded by a dielectric of sufficiently large ϵ/ϵ_0 and c/a. Of course, if

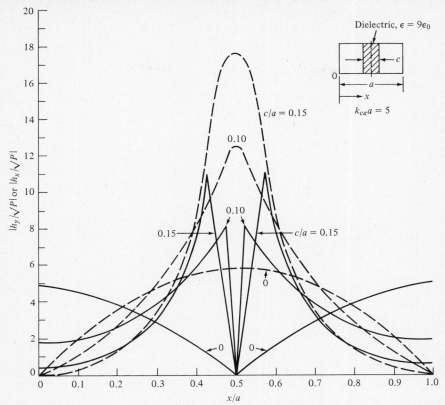

Figure 9.4 Magnetic field distribution in a dielectric-loaded guide. (Solid curves: $|h_y/\sqrt{P}|$; dashed curves: $|h_x/\sqrt{P}|$.)

c/a is too large, higher-order modes can be excited. It is then necessary to reduce the width of the guide in the dielectric loaded section in order to suppress them.

We further note from Fig. 9.4 that because of the concentration of electric field inside the dielectric, the magnetic field intensity is also intensified at the air-dielectric interfaces. This intensification is clearly advantageous for obtaining high values of isolation and phase shift.

In Fig. 9.5 the ellipticity e is plotted vs $k_{ca}a$. Solid lines represent e vs $k_{ca}a$ at the dielectric-air interfaces of a waveguide loaded by a centrally loaded dielectric, while the dashed line represents the ellipticity at x_1 of an empty guide. For the purpose of this plot, the value of x_1 has been arbitrarily chosen such that e is unity at $k_{ca}a = 4$. Note that the variation in e is greatly reduced by dielectric loading. As a consequence of this, substantially broader bandwidth can be expected. This fact, coupled with the magnetic field intensification near the air-dielectric interfaces should give rise to isolators with higher attenuation (and possibly higher reverse-to-forward attenuation ratios over a broader band-

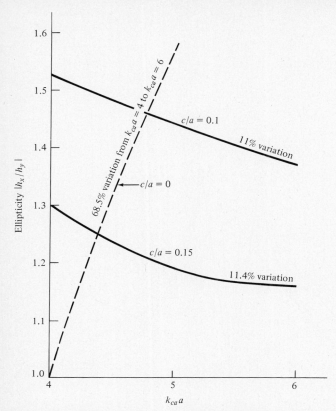

Figure 9.5 Ellipticity vs $k_{ca}a$. Solid lines: ellipticity vs $k_{ca}a$ at the dielectric-air interface of a waveguide loaded by a centrally located dielectric. Dashed line: the ellipticity at x_1 of an empty guide; the value of x_1 has been arbitrarily chosen such that the ellipticity is unity at $k_{ca}a = 4$.

width), as exemplified by typical characteristics of isolators with and without dielectric loading shown in Fig. 9.6. Coaxial isolators also use a form of dielectric loading.

9.3 PHASE SHIFTERS

It is clear that if the biased magnetic field of Fig. 9.1 is far away from resonance, reciprocal or nonreciprocal phase shift at low attenuation can be obtained. If nonreciprocal phase shift is desired, the field directions must be as indicated in Fig. 9.1. On the other hand, if reciprocal phase shift is needed, $\mathbf{H}_0$ must be applied in the same direction. Although electronically variable phase shifts can be obtained by varying the magnitude of H_0, not many of these phase shifters are in actual use.

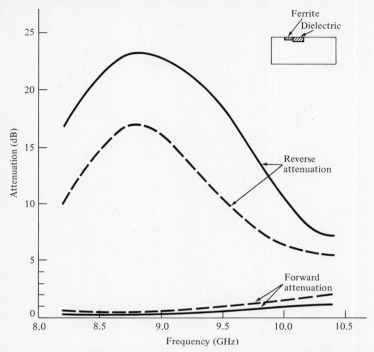

Figure 9.6 Attenuation vs frequency for a dielectric-loaded isolator (solid line) compared with the unloaded case (of ferrite only—the dashed line).

A more interesting and useful device is the latching ferrite phase shifter. It can be used, for example, as a digital phase shifter in changing the phase of antenna elements in an array for radar scanning purposes. Furthermore, the ferrite core used in this case is usually loaded by a dielectric insert; thus, the study of this device can further elucidate the effect of dielectric loading already discussed in Section 9.2.

Figure 9.7(a) shows the usual configuration of a latching or remanence phase shifter. A torroidal configuration is used to provide a closed magnetic circuit when the switching current I is applied. The phase shifter is nonreciprocal since the opposite directions of I give rise to a dc flux in the core that is oppositely directed. Furthermore, this differential phase shift can be varied by changing the magnitude of the current, if desired, as in the case of a scanning antenna array. To simplify the analysis, the idealized configuration of Fig. 9.7(b) is used.[6] Unlike the configuration of Fig. 9.7(a), a pure TE_{10} mode can be propagated in this latter case, greatly facilitating analytical calculation.

Latching ferrite phase shifters operated in the remanent state. If the sample is assumed to be lossless, the permeability components follows from Eq.

[6]W. J. Ince and E. Stern, *IEEE Trans. Microwave Theory Tech.* **MTT-15**, 87 (1967).

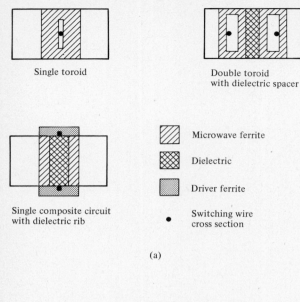

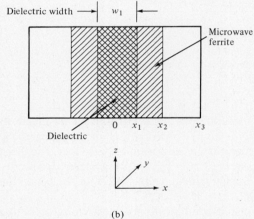

Figure 9.7 **(a)** Nonreciprocal remanence phase shifter configurations. **(b)** Simplified model of remanence phase shifter.

(7.26) as

$$\mu' = 1 + \chi' \simeq 1$$
$$\chi'' = 0$$
$$\kappa' = \frac{|\gamma| M_r}{\omega} \qquad (9.23)$$
$$\kappa'' = 0$$

where $M_r = \langle M_0 \cos \theta \rangle$ is the remanent magnetization in the ferrite core. Accordingly, the tensor permeability is given by

$$\|\boldsymbol{\mu}_\pm\| = \mu_0 \begin{bmatrix} 1 & \mp j \dfrac{|\gamma| M_r}{\omega} & 0 \\ \pm j \dfrac{|\gamma| M_r}{\omega} & 1 & 0 \\ 0 & 0 & 1 \end{bmatrix} \tag{9.24}$$

The remanent magnetization is considered positive if the r.f. magnetic field rotates in a clockwise direction about the direction of M_r. The form of Eq. (9.24) implicitly assumes that the average internal magnetic field is small and that, because of the magnitude of internal field, the number of domains in the remanent material that are at resonance is also small. Aside from eddy current and hysteresis losses, there are in addition domain wall relaxation and domain rotation resonance in ferrites.[7] Whereas domain relaxation occurs at the 50-MHz range, domain rotation resonance occurs at about 1 GHz. Thus, residual losses in remanent ferrites are due mainly to domain rotation resonance. It has been shown that the maximum frequency $\omega_{\max}$ at which domain rotation resonance could occur is[8,9]

$$\omega_{\max} = |\gamma|(H_a + M_0) \tag{9.25}$$

where H_a is the anisotropy field and M_0 is the saturation magnetization. Thus, it follows from Eq. (9.25) that to obtain small losses in the remanent state, it is necessary to satisfy the inequality

$$|\gamma| M_0 < \omega \tag{9.25a}$$

This general rule is found to be applicable to many ferrites used in microwave-latching phase shifters.

Let us now return to the calculation in connection with Fig. 9.7(b). The electric field distribution in regions 1 (dielectric), 2 (ferrite), and 3 (air) follows from Eqs. (3.11), (3.30), and (3.31):

$$E_{z1} = E_1 \cos(2\pi F k_1 x) \tag{9.26}$$

$$E_{z2} = E_2 e^{j2\pi F k_2 x} + E_4 e^{-j2\pi F k_2 x} \tag{9.27}$$

$$E_{z3} = E_3 \sin[2\pi F k_3 (x_3 - x)] \tag{9.28}$$

where we have made use of the fact that $E_{z3} = 0$ at $x = x_3$ and that the electric

[7]Soohoo, *Ferrites*, p. 83.

[8]D. Polder and J. Smit, *Rev. Mod. Phys.* **28**, 89 (1953).

[9]Soohoo, *Ferrites*, p. 86.

field is symmetric about the z axis. In these equations[10]

$$F = \frac{\omega}{\omega_c} \tag{9.29}$$

$$k_1 = \frac{k_{c1}}{2\pi F} \qquad k_2 = \frac{k_{c2}}{2\pi F} \qquad k_3 = \frac{k_{c3}}{2\pi F} \tag{9.30}$$

where ω_c is the center frequency. Thus, at $\omega = \omega_c$, k_1 is simply equal to $k_{c1}/2\pi$ where k_{c1} is the cutoff wave number in region 1.

Substituting Eqs. (9.26)–(9.28) into Eqs. (3.28) and (3.29), we find the magnetic field components as

$$H_{x1} = \eta E_1 p \cos(2\pi F k_1 x) \tag{9.31}$$

$$H_{y1} = j\eta E_1 k_1 \sin(2\pi F k_1 x) \tag{9.32}$$

$$H_{x2} = \frac{2\eta E}{(1 - m^2)} \{ p[r \cos(2\pi F k_2 x) - \sin(2\pi F k_2 x)]$$
$$\pm m k_2 [r \sin(2\pi F k_2 x) + \cos(2\pi F k_2 x)] \} \tag{9.33}$$

$$H_{y2} = \frac{j2\eta E}{(1 - m^2)} \{ \pm mp[r \cos(2\pi F k_2 x) - \sin(2\pi F k_2 x)]$$
$$+ k_2 [r \sin(2\pi F k_2 x) + \cos(2\pi F k_2 x)] \} \tag{9.34}$$

$$H_{x3} = \eta p E_3 \sin[2\pi F k_3 (x_3 - x)] \tag{9.35}$$

$$H_{y3} = j\eta E_3 [2\pi F k_3 (x_3 - x)] \tag{9.36}$$

where $\eta = \sqrt{\epsilon_0/\mu_0}$; p is the ratio of free space to guide wavelength, λ/λ_g or $2\pi p/\lambda = \beta$; and $m = |\gamma| M_r/\omega$. E and r are newly introduced constants.

In order for Eqs. (9.26)–(9.28) and (9.31)–(9.36) to be self-consistent, as in the pure-dielectric-loading case of Eqs. (9.17) and (9.18), a set of three simultaneous equations containing p must be satisfied:

$$k_1^2 = \epsilon_1' - p^2 \tag{9.37}$$

$$k_2^2 = \epsilon_2'(1 - m^2) - p^2 \tag{9.38}$$

$$k_3^2 = \epsilon_3' - p^2 \tag{9.39}$$

where $\epsilon_1' = \epsilon_1/\epsilon_0$, $\epsilon_2' = \epsilon_2/\epsilon_0$, and $\epsilon_3' = \epsilon_3/\epsilon_0$.

Applying the boundary conditions of the continuity of the electric field at the dielectric-ferrite and ferrite-air interfaces, we have

$$E_1 \cos(2\pi F k_1 x_1) = E_2 e^{j2\pi F k_2 x_1} + E_4 e^{-j2\pi F k_2 x_1} \tag{9.40}$$

$$E_3 \sin[2\pi F k_3 (x_3 - x_2)] = E_2 e^{j2\pi F k_2 x_2} + E_4 e^{-j\pi F k_2 x_2} \tag{9.41}$$

---

[10]Here we have used the notations of Ince and Stern (footnote 6). Note that they used normalized quantities.

Solving Eqs. (9.37)–(9.39) simultaneously with the aid of Eqs. (9.40) and (9.41) finally yields the characteristic equation

$$\frac{\tan(\theta_4 - \theta_2)}{k_2}$$

$$= \frac{k_3 \cot \theta_3 - k_1 \tan \theta_1}{(\epsilon_2' - p^2) + (1 - m^2)k_1 k_3 \tan \theta_1 \cot \theta_3 \pm mp(k_3 \cot \theta_3 + k_1 \tan \theta_1)} \tag{9.42}$$

where

$$\begin{aligned}
\theta_1 &= 2\pi F k_1 x_1 \\
\theta_4 - \theta_2 &= 2\pi F k_2 (x_2 - x_1) \\
\theta_3 &= 2\pi F k_3 (x_3 - x_2) \\
\theta_4 &= 2\pi F k_2 x_2
\end{aligned} \tag{9.43}$$

Numerical solution of Eq. (9.42) will yield field values of p for specific combinations of parameters. There are two solutions, p^+ and p^-, corresponding to clockwise and counterclockwise orientation of the remanent flux about the direction of propagation in the idealized twin-slab phase shifter of Fig. 9.7(b). A plot of p^+ and p^- vs the normalized magnetization $m = |\gamma| M_r / \omega_0$ is shown in Fig. 9.8. For the purpose of this computation $\omega_0 = 10$ GHz and the waveguide dimensions are those of a standard X-band waveguide. Note that p^+ and p^- are degenerate at $m = 0$ as they should since this case corresponds to pure dielectric loading. Furthermore, whereas p^+ decreases, p^- increases with increasing m; this behavior is consistent with the dependence of μ_+' and μ_-' on H_z shown

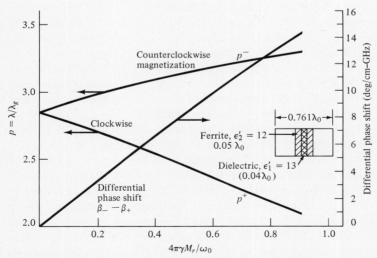

Figure 9.8 Normalized waveguide wavelength and differential phase shift vs normalized remanence magnetization.

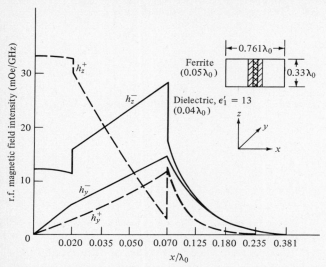

Figure 9.9 Magnetic field intensity for 1 W of incident power. $(4\pi\gamma M_r/\omega_0 = 0.8.$ λ_0 is the free-space wavelength at $\omega_0.$ Dashed curves: clockwise magnetization; solid curves: counterclockwise magnetization.)

in Fig. 8.2. The differential phase shift per unit length, the quantity we seek, follows from the definition $(2\pi/\lambda)p = \beta$:

$$\beta_- - \beta_+ = \frac{2\pi}{\lambda}(p^- - p^+) \tag{9.44}$$

This quantity is also plotted in Fig. 9.8. Note that $\beta_- - \beta_+$ increases nearly linearly with m. Furthermore, quite large values of $\beta_- - \beta_+$ are obtained as $m \to 1$. However, as $m \to 0$, remanent state losses increase [see inequality (9.25a)], which is of course undesirable. Also, as $\beta_- - \beta_+$ increases, it becomes more difficult to obtain a good match for both states of magnetization. Conceptually, it is clear that $\beta_- - \beta_+$ is a function of all parameters of the devices, $m, \epsilon_1, \epsilon_2, \epsilon_3$, the thicknesses of the regions in Fig. 9.7, as well as the operating frequency ω. For details of the dependence of $\beta_- - \beta_+$ on these parameters, the reader is referred to the paper by Ince and Stern.[11]

Once the eigenvalues of p are found by a numerical solution of Eq. (9.42), the field components (eigenfuctions) can be found from Eqs. (9.26)–(9.28) and (9.31)–(9.36). Figure 9.9 shows the r.f. magnetic field intensity vs x/λ_0 for $m = 0.8$ and unit incident power. For clarity, the abscissa scale has been changed at the ferrite-air interface. Within much of the ferrite the magnetic field is linearly polarized with a small component of circular polarization. The energy distribution in the dielectric and ferrite regions differs widely for the two directions of propagation. This may explain, in part, the effectiveness of dielectric loading in phase shifters; the ferrite tends to displace the field outside the dielectric

[11]Ince and Stern, *IEEE Trans.*

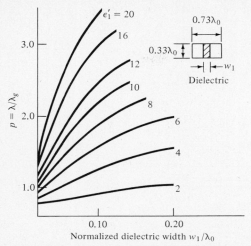

Figure 9.10 Normalized waveguide wavelength vs normalized dielectric width for a dielectric-loaded waveguide.

region for the negative wave; other things being constant, this field displacement increases with ϵ_1. The maximum magnetic field intensity in the ferrite occurs in the region adjacent to the ferrite-dielectric interface. However, the magnitude of the magnetic field in the ferrite is not affected to any great extent by dielectric loading.

In a practical phase shifter a matching network is required to match the phase shifter section to the feeding waveguide. One convenient type of matching structure is the n-section Tchebyscheff transformer.[12] The basic design parameter λ_g of each section may be obtained from the characteristic equation (9.42) by reducing the ferrite thickness to zero. Putting $\theta_2 = \theta_4$ in Eq. (9.42), we readily simplify it to

$$k_1 \cot \theta_1 = k_3 \cot \theta_3 \qquad (9.45)$$

Some results from a numerical solution of Eq. (9.45) are given in Fig. 9.10. Here p is plotted against the normalized dielectric width w_1/λ_0 with ϵ_1' as a constant parameter.

To complete our analysis, we need to compute the loss/length of the phase shifter. This can be done by assuming that the field distribution is not appreciably affected by the presence of these losses, dielectric, magnetic, and wall. However, the computation is rather laborious. Figure 9.11 shows the decibel loss per 2π differential phase shift as a function of frequency. Note that these losses are relatively low.

Before we conclude our discussion on remanent ferrite phase shifters, let us recall that the idealized model of Fig. 9.7(b) differs in several respects

[12]G. L. Matthaei et al., *Microwave Filters, Impedance-Matching Networks and Coupling Structures*, McGraw-Hill, New York, 1964, Ch. 6.

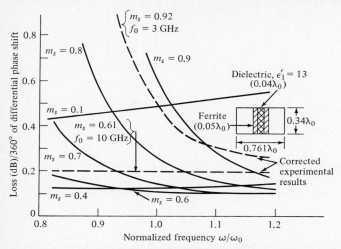

Figure 9.11 Decibel loss per 360° differential phase shift vs normalized frequency.

from the configuration of a real phase shifter as depicted in Fig. 9.7(a). The connecting members that complete the magnetic path of the toroid are not accounted for in the model, the flux in the connecting members is parallel to the broad walls of the waveguides, and there are leakage flux effects. It was found that the corners of toroids contributed little to the observed differential phase shift. Nevertheless, the theoretical and experimental results on phase shift and attenuation of the device seem to agree reasonably well if appropriate correction factors are applied for the deviations from the ideal model that have been mentioned.[13]

9.4 CIRCULATORS

A *circulator* is a microwave ferrite component with three or four ports. Within the circulator, energy can propagate with the circulation pattern of increasing port numbers or vice versa depending on the direction of the dc field applied to the ferrite. Thus, in the schematic diagram of a three-port circulator shown in Figure 9.12(a), energy circulation can be in the $1 \to 2 \to 3$ or $1 \to 3 \to 2$ directions. In the former case all energy fed into port 1 will emerge from port 2 and none from port 3. For this to occur, the clockwise and counterclockwise traveling wave components arriving at 2 must differ in phase by an integral multiple of 2π, while those arriving at 3 must differ by an odd multiple of π. Thus, for a perfect H-plane junction circulator, we must have

$$(2\beta_- l + 2 \Delta\phi_-) - (\beta_+ l + \Delta\phi_+) = m2\pi \tag{9.46}$$

$$-(\beta_- l + \Delta\phi_-) + (2\beta_+ l + 2\Delta\phi_+) = (2n - 1)\pi \tag{9.47}$$

[13]Ince and Stern, *IEEE Trans.*

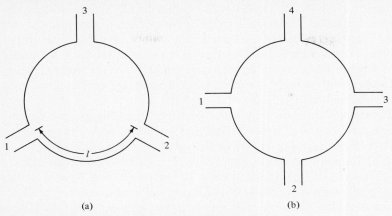

(a)   (b)

Figure 9.12   **(a)** Three- and **(b)** four-part circulators.

where m and n are positive or negative integers including zero. Here β_+ and β_- are phase constants for waves traveling in the counterclockwise and clockwise directions, respectively, while $\Delta\phi_+$ and $\Delta\phi_-$ are additional phase shifts due to nonpropagating higher-order modes excited at each of the junctions corresponding to counterclockwise and clockwise propagations, respectively. Solving Eqs. (9.46) and (9.47) simultaneously, we find

$$\beta_+ l = \frac{2m + 4n - 2}{3}\pi - \Delta\phi_+ \tag{9.48}$$

$$\beta_- l = \frac{4m + 2n - 1}{3}\pi - \Delta\phi_- \tag{9.49}$$

and

$$\beta_- l - \beta_+ l = \frac{2m - 2n + 1}{3}\pi + (\Delta\phi_+ - \Delta\phi_-) \tag{9.50}$$

For example, if $m = n = 1$, $\beta_+ l = 4\pi/3 - \Delta\phi_+$, $\beta_- l = 5\pi/3 - \Delta\phi_-$, and $\beta_- l - \beta_+ l = \pi/3 + (\Delta\phi_+ - \Delta\phi_-)$. Note that $\Delta\phi_+$ and $\Delta\phi_-$ depend on the geometry of the junction as well as on the location and geometry of the ferrite. The same reasoning can be applied when power enters ports 2 or 3 instead. Since the number designation of the ports is entirely arbitrary, perfect circulation is insured if Eqs. (9.46)–(9.50) are satisfied.

For an E-plane junction circulator, Eqs. (9.46)–(9.49) are still applicable but $\Delta\phi_+$ and $\Delta\phi_-$ will take on different values compared to those for H-plane junction circulators.[14] The above phase shift requirements are for counterclockwise energy circulation. If $\mathbf{H}_0$ is reversed in direction, $\beta_+ \leftrightarrow \beta_-$ and the circulation will be clockwise instead.

---

[14]For an H-plane circulator the wave entering the side arm splits between the collinear arms with equal phase. Conversely, if waves in the collinear arms arrive at the junction with equal phase, they will combine in the side arm also with equal phase. On the other hand, for an E-plane junction, π relative phase shifts occur for the component waves of both the situations mentioned above. Consequently, these additional π-phase shifts need not be explicitly included in Eqs. (9.46)–(9.50).

Note that Eqs. (9.46) and (9.47) were formulated with the implicit assumption that power entering the input terminal divides equally between the two branch arms. This will indeed be the case if all terminals, or at least the output terminals, are correctly terminated in which case the impedances looking into the right or left branch arms of the input are equal. Since the number designation of the ports are entirely arbitrary, it is clear that for perfect circulation all ports must be matched.

To synthesize a three-port circulator obeying Eq. (9.50), we first observe that the phase shifts are assumed distributed and that both clockwise and counterclockwise propagation occurs in the junction proper. Clearly, then, a possible configuration is shown in Fig. 9.13(a). For definiteness, let us assume that the junctions are H-plane junctions with the ferrite ring located at the plane (cylindrical) of circular polarization. The direction of circulation is determined by the direction of the magnetizing field $\mathbf{H}_0$, into or out of the paper. In any event the component waves traveling in the opposite directions in the

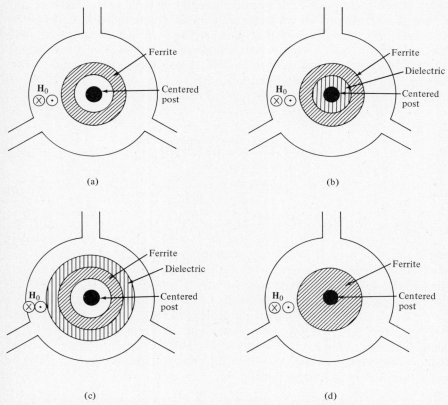

Figure 9.13   Three-port circulator ferrite-dielectric configurations. Note that H_0 can be directed either in or out of the page.

junction proper will have the opposite sense of polarization, giving rise to the required differential phase shift at the appropriate field H_0.

From our experience in isolator design we may expect the bandwidth of the circulator to increase if the ferrite ring is loaded by dielectric sleeves with radii smaller or larger than that of the ferrite sleeve as shown in Fig. 9.13(b, c). If the circulator is to carry high power, we can dispense with the inner dielectric ring of Fig. 9.13(b) and reduce the inner radius of the ferrite ring to that of the metallic post. In this way the heat dissipated in the ferrite will escape through the centered post. Such a resulting structure is shown in Fig. 9.13(d). Alternatively, if the inner radius of the ferrite ring of Fig. 9.13(a) can be reduced to that of the metallic post, it gives rise to the higher power structure of Fig. 9.13(d).

If the radius of the metallic post and the inner radius of the ferrite sleeve of Fig. 9.13(d) are simultaneously reduced to zero, we arrive at the configuration of the ferrite-post circulator. In this case we would not expect the r.f. magnetic field to be circularly polarized throughout the ferrite, at least not over an extended frequency range. Indeed, it is well known that this type of circulator has rather narrow bandwidth. However, it has the obvious advantage of simplicity from both the theoretical and practical standpoints. For this reason its detailed analysis will be given in Section 9.5.

For the four-port circulator of Fig. 9.12(b), one may be tempted to write a set of phase shift equations similar to Eqs. (9.46) and (9.47). However, it turns out that perfect circulation cannot be obtained with the simple E-plane or H-plane junction structure of Fig. 9.12(b) (although this figure is still useful as a schematic representation of a four-port circulator). It is left as an exercise for the reader to show that this is indeed the case (see Problem 9.3).

Conceptually, a circulator can be constructed with two *magic tees*, two-phase shifters with a differential phase shift of 90°, and a 90° reciprocal phase shifter, connected by sections of waveguides, as shown in Fig. 9.14(a). Power entering terminal 1 will split equally between b and d.[15] These component waves will arrive at a and c in equal phase, provided the physical lengths d-c and b-a are equal. They will thus combine and appear at terminal 2.[16] Similarly, power entering arm 2 will appear only in arm 3, while power entering arm 3 will only appear in arm 4. Thus, perfect circulation is obtained. If the magnetizing field H_0 of the ferrite differential phase shifter is reversed, it can be shown in a similar fashion that the sequence of circulation is reversed from $1 \rightarrow 2 \rightarrow 3 \rightarrow 4 \rightarrow 1$ to $1 \rightarrow 4 \rightarrow 3 \rightarrow 2 \rightarrow 1$.

A more compact four-port ferrite circulator is shown in Fig. 9.14(b).[17] In this configuration two short-slot hybrids or directional couplers are used in conjunction with phase shifters. As shown, the insertion of the dielectric causes

[15] Power entering 1 will not appear in arm 3 because TM modes are usually below cutoff at the operating frequency.

[16] Note that a-c-2 constitutes an H-plane junction while a-c-4 an E-plane junction. Accordingly, waves arriving at a and c in phase combine in phase in arm 2 but out of phase in arm 4.

[17] R. F. Soohoo, *Ferrites*, p. 179.

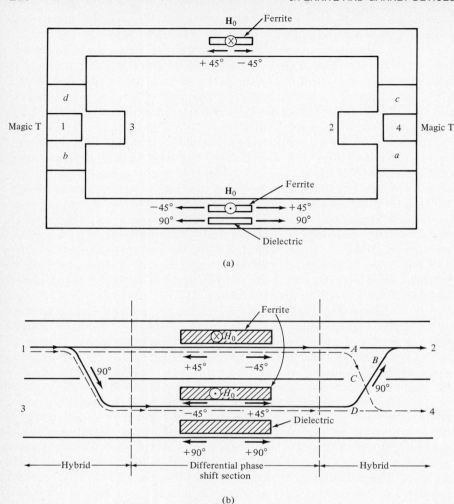

Figure 9.14   Two configurations of transverse-field circulators.

an additional phase shift of 90° in the waveguide. On the other hand the phase shifts for the ferrite sections are relative to the phase shifts at zero biasing field and are not additional phase shifts due to the insertion of the magnetized ferrite. This is permissible if the ferrite sections are identical in geometry.

Power entering terminal 1 is split into two equal parts, one part traveling in the upper guide toward terminal 2 while the other part goes to the lower guide (undergoing a 90° phase shift in the hybrid) and travels toward terminal 4. If we add up the phase shifts of paths A and B, we find that they are $-45°$ and 315° and are therefore in phase, while the phase shifts for path C and D are 45° and 225° and are thus opposite in phase. Thus, power entering 1 will emerge from 2. Similarly power entering 2 will emerge from 3, power entering

3 will emerge from 4, and so on. If **H** is changed in direction, the circulation sequence will change from $1 \to 2 \to 3 \to 4 \to 1$ to $1 \to 4 \to 3 \to 2 \to 1$, as
expected.

9.5 ANALYSIS OF JUNCTION CIRCULATORS

From a synthesis standpoint there is no conceptual difference between a waveguide circulator, a stripline circulator, and a coaxial circulator. Of course, this
underlying unity does not preclude differences in detail in the synthesis pro-
cedures or in operating characteristics for circulators using different waveguiding structures.

There are basically two ways to approach a synthesis problem, namely,
the equivalent network approach and the scattering matrix approach. In the
former case a lumped-element network having the desired characteristics is first
obtained by conventional synthesis technique, and then a microwave structure
exhibiting the same characteristics is synthesized by interconnecting basic waveguide components, represented by lumped-element networks and sections of
transmission lines. At microwave frequencies the physical structure of the circuit
is not defined by the network diagram as at lower frequencies. For this reason
there is often no particular advantage in working from an explicit network
representation of the desired circuit in the case of junction circulators.

In the second synthesis method the microwave circuit is considered as a
waveguide junction characterized by a scattering matrix S without reference to
a specific network representation. The scattering matrix coefficients are then
fixed in accordance with the desired circuit characteristics by suitably adjusting
the geometry of the boundary value problem. This is a particularly useful
approach when the microwave circuit has structural symmetry, in which case it
is convenient to regard the eigenvectors and eigenvalues of S rather than the
scattering coefficients themselves as adjustable parameters of the circuit. Since
a junction circulator clearly possesses a certain degree of symmetry, it is indeed
convenient to use the scattering matrix approach to its synthesis.

Consider first the case of an m-port symmetrical H-plane waveguide with
a central ferrite post as shown in Fig. 9.15. We shall match exactly each
cylindrical mode in the ferrite post to the associated mode outside the post,
which in turn is matched exactly to the complete set of modes of the rectangular
guide.[18] An infinity of equations results (one equation for each cylindrical
mode), each involving an infinity of unknowns (amplitudes of the rectangular
waveguide modes). One then develops in detail an approximation to this infinite
linear system that is equivalent to ignoring the evanescent modes of the rectangular waveguide. By considering solutions that correspond to each eigen-
vector in terms of the scattering matrix, one need match only the fields of one
of the m waveguide modes.

[18]J. B. Davies, *IRE Trans. Microwave Theory Tech.* **MTT-10,** 596 (1962)

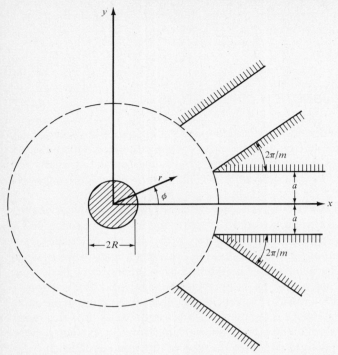

Figure 9.15 Configuration of ferrite post in the m-port waveguide junction.

9.5.1 The Scattering Matrix[19]

Let the waveguide terminal planes be equidistant from the junction axis, this distance being half an integral number of guide wavelengths. These planes are sufficiently far from the junction that higher-order modes in the waveguide excited at the junction have decayed to negligible values. Thus, fields are due only to the propagating waveguide modes. Let a_i be the electric field of the incident wave at the terminal plane of the ith port with the ports consecutively numbered around the junction. Similarly, let b_i be the electric field of the outgoing wave.

If **a** and **b** are column vectors with elements a_i and b_i, respectively, then the relation $\mathbf{b} = [S]\mathbf{a}$ defines the scattering matrix $[S]$. From the symmetry property of a lossless, matched, nonreciprocal three-port junction, it can be shown that $[S]$ must have the form[20]

$$[S] = \begin{bmatrix} 0 & S_{12} & S_{13} \\ S_{21} & 0 & S_{23} \\ S_{31} & S_{32} & 0 \end{bmatrix} \qquad (9.51)$$

[19]C. G. Montgomery, R. H. Dicke, and E. M. Purcell, *Principles of Microwave Circuits*, McGraw-Hill, New York, 1948, Chs. 5 and 12.

[20]H. J. Carlin, *Principles of Gyrator Networks*, Polytechnic Institute of Brooklyn, Microwave Res. Inst. Symp. Ser., Vol. 4, 1955, p. 175.

where the element S_{ij} represents the amplitude of the outgoing wave at terminal i as a result of a wave of unit amplitude incident at terminal j. For a reciprocal junction the scattering matrix is symmetrical; i.e., $S_{ij} = S_{ji}$. For a nonreciprocal junction, however, $[S]$ is no longer symmetrical, so that $S_{ij} \neq S_{ji}$. Nevertheless, if the junction is lossless, conservation of power still requires that $[S]$ be unitary; i.e., $[S][\tilde{S}]^* = 1$. Application of this condition to Eq. (9.51) yields

$$S_{12}S_{12}^* + S_{13}S_{13}^* = 1 \qquad (9.52a)$$

$$S_{21}S_{21}^* + S_{23}S_{23}^* = 1 \qquad (9.52b)$$

$$S_{31}S_{31}^* + S_{32}S_{32}^* = 1 \qquad (9.52c)$$

$$S_{12}S_{32}^* = S_{23}S_{13}^* = S_{31}S_{21}^* = 0 \qquad (9.52d)$$

$$S_{12}^*S_{32} = S_{23}^*S_{13} = S_{31}^*S_{21} = 0 \qquad (9.52e)$$

Either one of the last two expressions can be considered redundant. Let us assume that $S_{21} \neq 0$. Then Eq. (9.52e) gives $S_{31}^* = 0$. Equation (9.52c) now requires $|S_{32}| = 1$, and thus $S_{12}^* = 0$ from Eq. (9.52e) and $|S_{13}| = 1$ from Eq. (9.52a). Thus, we see that $|S_{21}| = 1$, also from Eq. (9.52b). Thus, $[S]$ of Eq. (9.51) simplifies to

$$[S] = \begin{bmatrix} 0 & 0 & S_{13} \\ S_{21} & 0 & 0 \\ 0 & S_{32} & 0 \end{bmatrix} \qquad (9.53)$$

indicating circulation in the sequence $1 \rightarrow 2 \rightarrow 3 \rightarrow 1$. Since the output waveguides are symmetrically located, the numbering of the ports, although necessarily consecutive, is entirely arbitrary. If the location of the terminal planes is properly chosen, the phase angles of S_{13}, S_{21}, and S_{32} can be made zero. In that case $S_{13} = S_{21} = S_{32} = 1$. If suitable tuning elements are placed in each arm (these can be identical because of the threefold junction symmetry), the junction can be matched so that $S_{11} = S_{22} = S_{33}$ as implied in matrix (9.51). In the analysis above we found that perfect circulation occurs for a lossless junction. In an actual circulator junction losses are, of course, unavoidable. As a consequence, transmission losses, although small, do occur. Correspondingly, isolation between ports is also finite.

9.5.2 Field Analysis

The field analysis of an m-port junction with a central ferrite post extending across the full height of the junction is rather involved. For this reason the reader is referred elsewhere[21] for details of the analysis. Here we shall concentrate on the conceptual approach to the solution of the problem and to discuss the results obtained.

[21]Davies, *IRE Trans.*, p. 596.

Refer again to Fig. 9.15. Since fields throughout the junction are excited by TE_{10} modes at the waveguide terminal planes, and as there are no z-dependent boundary conditions, we look for solutions with electric fields purely in the z direction and magnetic fields purely in the x-y plane that are independent of z. A complete expansion for fields inside the ferrite ($r \leq R$) can then be written. Similarly, fields outside the ferrite ($r \geq R$) can also be written in series form. Imposing the continuity boundary conditions on E_z and H_ϕ at $r = R$ to the individual modes of the series above, we find

$$\frac{C_n}{B_n} = -\frac{J_n(\gamma_0 R)}{Y_n(\gamma_0 R)} \left\{ \left[\frac{J_n'(\gamma R)}{\gamma R J_n(\gamma R)} + \frac{\kappa n}{\mu(\gamma R)^2} \right] \frac{\epsilon}{\epsilon_0} - \frac{J_n'(\gamma_0 R)}{\gamma_0 R J_n(\gamma_0 R)} \right\}$$

$$\times \left\{ \left[\frac{J_n'(\gamma R)}{\gamma R J_n(\gamma R)} + \frac{\kappa n}{\mu(\gamma R)^2} \right] \frac{\epsilon}{\epsilon_0} - \frac{Y_n'(\gamma_0 R)}{\gamma_0 R Y_n(\gamma_0 R)} \right\}^{-1} \tag{9.54}$$

where C_n and B_n are coefficients in the expansion for E_z in the region outside the ferrite post ($r \geq R$), i.e., in the expresion

$$E_z = \sum_{n=-\infty}^{\infty} [B_n J_n(\gamma_0 r) + C_n Y_n(\gamma_0 r)] e^{-jn\phi} \tag{9.55}$$

Also, ϵ is the ferrite dielectric constant, while μ and κ are elements of its permeability tensor:

$$\|\boldsymbol{\mu_r}\| = \begin{bmatrix} \mu & -j\kappa & 0 \\ j\kappa & \mu & 0 \\ 0 & 0 & 1 \end{bmatrix} \tag{9.56}$$

$\gamma_0^2 = \omega^2 \mu_0 \epsilon_0$, $\gamma^2 = \omega^2 \epsilon (\mu^2 - \kappa^2)/\mu$ and ω is the frequency.

Next a complete expansion for the fields in the waveguide can be written. Then, imposing the continuity boundary condition on E_z and H_ϕ along the dotted boundary of Fig. 9.15, we finally obtain a set of infinite linear equations with an infinity of unknown coefficients D_p, the waveguide mode amplitude. This infinite linear system, culminating in the limit G_{-1}/G_1 with finite determinants, is a precise exact formulation of the junction problem. To solve this set of equations, it is clearly necessary to resort to approximate methods. Successive approximate solutions to the infinite system can be derived from one of the linear equations, from three equations, and so on. Compared with the exact analysis, the adopted approximation is equivalent to ignoring the evanescent modes of the rectangular waveguide arms while no approximation is made on the cylindrical modes. In any event the approximate result takes the form

$$\tan[\tfrac{1}{2}(\theta_j + \pi)] = \frac{D_n^m + E_n^m(C_n/B_n)}{F_n^m + G_n^m(C_n/B_n)} \tag{9.57}$$

where C_n/B_n is a function only of the ferrite parameters and radius as given by Eq. (9.54) while D_n^m, E_n^m, F_n^m, and G_n^m are functions only of the waveguide parameters (m_y and λ_0/a). θ_j is defined by the equation $\lambda_j = e^{j\theta_j}$ with $\lambda_j = G_{-1}/G_1$.

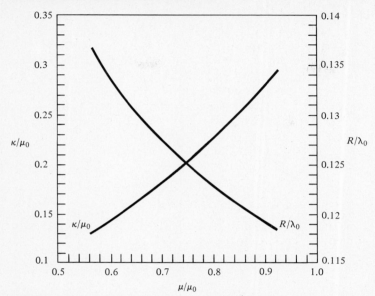

Figure 9.16   Relationship of μ, κ, and R for perfect circulation in a three-port Y circulator.

Solving Eq. (9.57) by numerical methods, the approximate scattering matrix for a given set of ferrite and junction parameters at any given frequency can be obtained. For a three-port junction perfect circulation is predicted over a small range of ferrite post radius, and the relation between μ, κ, and ferrite radius for such circulation is given in Fig. 9.16. If we superimpose on the figure a curve of practical κ vs μ for some typical ferrite, the intersection with the theoretical curve will give the appropriate μ and κ, as well as ferrite post diameter to be used. Experimentally, it is found that the isolation, VSWR, and insertion loss can vary greatly with R (see Fig. 9.17). Note that the theoretical value of R is only slightly larger than the optimum experimental value.

For the three-port junction where two conditions have to be satisfied, we find a continuous (but narrow) range of μ, κ, and R values consistent with circulation. For the four-port junction none of the μ, κ, and R values computed gives rise to a circulator,[22] and therefore a convenient fourth variable has to be introduced. In place of the central post of full-height ferrite, a post in sandwich form consisting of many thin ferrite disks interspaced by thin conducting disks can be used. If it is assumed that the ferrite disks are so thin as to support no propagating modes with the z dependence, it is theoretically found that whereas no perfect four-port circulator is possible, a number of "near-circulators" are predicted with the values of μ and κ used in practice.

[22]This is consistent with the results of Problem 9.3.

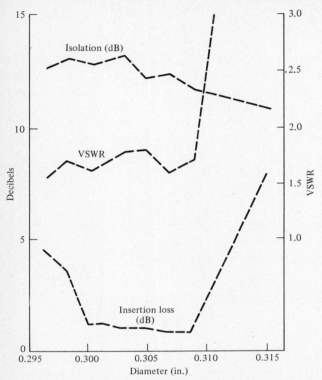

Figure 9.17 Experimental performance of a Y circulator with Ferroxcube D5 at 9273 MHz, with a predicted optimum diameter of 0.314 in.

Subsequent to Davies's analysis[23] outlined here, Costillo and Davis applied Davies's method to various inhomogeneous cylindrical ferrite structures including the ferrite tube–dielectric rod–dielectric sleeve and ferrite post–metal pin–dielectric sleeve structures that extend the full height of the waveguide.[24] Qualitative agreement between theory and experiment was obtained.

The most important papers on the synthesis of stripline circulators appear to have been those of Bosma[25] and Davies and Cohen.[26] In Bosma's first paper,[27] using simplified boundary conditions, a Fourier analysis produced two equations governing circulations that would normally require numerical solutions. Unfortunately, Bosma used simplified approximations that produced invalid analytical expressions in explicit form. In his subsequent paper[28] the

[23]Davies, *IRE Trans.*, p. 296.

[24]J. B. Costillo, Jr. and L. E. Davis, *IEEE Trans. Microwave Theory Tech.* **MTT-18,** 25 (1970).

[25]H. Bosma, *Proc. Inst. Electrical Engrs.* **109**, Pt. B, Suppl. 21, 137–146 (1962); *IEEE Trans. Microwave Theory Tech.* **MTT-12**, 61 (1964).

[26]J. B. Davies and P. Cohen, *IEEE Trans. Microwave Theory Tech.* **MTT-11**, 506 (1963).

[27]Bosma, *Proc. Inst. Elec. Engr.*

[28]Bosma, *IEEE Trans.*

problem was reformulated in terms of a Green's function. Equations were then obtained that are more amenable to approximations. Valid solutions were obtained that agree well with experiment. These approximations amount to considering only one of the many possible modes of circulation and restricting the applied field to values that produce small ratio of κ/μ. The Davies and Cohen paper[29] helped to establish the validity of the Bosma circulation equations and to extend their validity to all values of μ and κ by deriving the scattering matrix for the junction. Since the approach of Davies and Cohen has more practical applicability (i.e., to all values of μ and κ), and since the scattering matrix approach for the stripline circulator is quite akin to that for a waveguide circulator already discussed, further discussion of the subject will not yield additional conceptual insight. For more details, the reader is referred elsewhere.[30,31]

9.6 APPLICATIONS

Isolators, phase shifters, and circulators are widely used in microwave systems for input-output isolation, switching, etc. In what follows we shall briefly describe the chief applications for each component.

9.6.1 Isolator Applications

Ferrite isolators are routinely used to isolate source and load. This isolation helps to improve the frequency stability of, for example, a reflex Klystron oscillator whose frequency of oscillation is affected by the load impedance.[32] Before the advent of ferrite isolators, an attenuator (~ 6 dB) was placed between the reflex Klystron and the load to absorb the reflected energy. Unfortunately, this attenuator, being bilateral, attenuates the power traveling from source to load just as well. Indeed, with a 6-dB pad, 75% of the source power is dissipated in the attenuator. If an isolator is used in place of the attenuator, the power loss in the forward (oscillator-to-load) direction can be quite small, while that in the reverse (load-to-oscillator) direction can be sufficiently large to ensure frequency stability.

To demonstrate the load power gain obtained in replacing an attenuator by an isolator, let us refer to the typical example illustrated in Fig. 9.18. In Fig. 9.18(a) the power supplied to the load, P_2, is only 25% of P_1 (i.e., P_2 is -6 dB down from P_1) owing to the insertion of the attenuator. Any reflections from the load are reduced by another 6 dB or to 6.25% of P_1. On the other hand the power P'_2 reaching the load in Fig. 9.18(b), where an isolator has been substituted for the attenuator, is 93.3% (-0.3 dB) of P_1. Any reflection from the

[29]Davies and Cohen, *IEEE Trans.*

[30]Bosma, *Proc. Inst. Elec. Engr.*; *IEEE Trans.*

[31]Davies and Cohen, *IEEE Trans.*

[32]R. F. Soohoo, *Microwave Electronics*, Addison-Wesley, Reading, MA, 1971, p. 95.

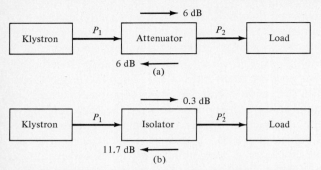

Figure 9.18   Load isolation using attenuator and isolator.

load would be attenuated another 11.7 dB, giving a round-trip attenuator of 12 dB. Thus, as far as the effect of the load mismatch is concerned, the reflex Klystrons of Fig. 9.18(a) and Fig. 9.18(b) are equally stable. If we were to define a figure of merit or transmission efficiency of the attenuator and isolator as

$$n_T = \frac{P_2}{P_1} \quad \text{or} \quad \frac{P_2'}{P_1}$$

we would have $n_T = 25\%$ for the attenuator and 93.3% for the isolator.

Most magnetrons are designed for optimum performance into a matched load or one in which the degree of mismatch is very minor. These idealized matching conditions are rarely present when a line many wavelengths long separates the magnetron from the antenna, owing to changes in the antenna match and the electrical line length as the frequency is varied.[33] This variation in load impedance with tuning can cause variations in the magnetron power output. The isolator, however, presents a good match to the magnetron over a wide frequency range. This enables the magnetron output power to be maintained at its maximum value at any given frequency. It also eliminates the tendency of a magnetron to lock on some frequencies or fail to operate on others. Indeed, pulsed magnetrons with mismatched loads may also transmit pulses of more than one frequency.[34]

The width of the frequency spectrum of a pulsed magnetron is not necessarily constant. When magnetrons are operated into a long line with a mismatched load, the frequency spectrum may tend to broaden and to vary for different pulses. A closer approach to theoretical spectrum is obtained when an isolator is used. Variation in impedance for the duration of a transmitted pulse may also be caused by the TR (transmit-receive) tube in a radar duplexing system. This source of frequency pulling may also be eliminated by the isolator.

Isolators can also be used to isolate one subsystem from another as well as to isolate source from a load. Referring to Fig. 9.19, any signal from sub-

---

[33]This impedance variation with frequency is evident from an examination of the expression for the input impedance Z_{in} given by Eq. (3.138).

[34]W. L. Pritchard, *Trans. IRE Microwave Theory Tech.* **MTT-4**, 97 (1956).

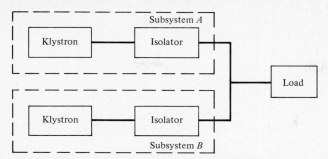

Figure 9.19 Use of isolators to isolate subsystems.

system A, for example, is attenuated by at least an amount in decibels equal to the sum of the forward attenuation of isolator A and reverse attenuation of isolator B before it reaches the Klystron of subsystem B. Thus, not only is the load isolated from subsystems A and B, but subsystems A and B are themselves isolated from each other.

An isolator will also protect a magnetron from high reflected power due to arching or component mismatch. Without an isolator, the high reflected power can seriously damage or destroy a magnetron.

For precise microwave measurements a stable, bilaterally matched system is needed. As we have explained, an isolator can increase oscillator stability by greatly attenuating the power reflected from the load. Such discussion implicitly assumes that the isolator is well matched for both directions of propagation. The use of such isolators thus increases not only system stability but also minimizes errors caused by multiple reflections. Minimization of such multiple reflection errors can also be accomplished by placing isolators in legs of these measuring bridges and interferometers.

Isolation can also be obtained by using a three-port junction circulator with one of its ports terminated. Thus, power entering port 1 (matched) will emerge from port 2 with little attenuation to enter the load. Reflections from the load will travel toward port 3 and into a matching termination. The source at port 1 will therefore be isolated from the reflections of the load connected to port 2. More detailed discussion of circulator application will be given in the next section.

9.6.2 Circulator Applications

Circulators with three or more ports can be used in many microwave applications, including isolation, duplexing, and digital phase shifting. They can be of the nonswitchable type (with ferrite biased by a permanent magnet) or switchable (biased by an electromagnet). They can be built in waveguides or in coaxial lines. Those most commonly used are of the junction rather than the differential phase shift type; the former has the advantage of smaller size, lower cost, and higher performance. Whereas the waveguide junction circulator is suitable for low- and medium-power applications, the waveguide differential phase-shift type is useful for high power. Coaxial junction circulators are fabricated in stripline with a transformation to coaxial connectors. Both coaxial types are used in low- and high-power circuits.

Although single-junction, four-port circulators have been built, they are inherently narrow-band and are not commonly used.[35] Instead, two three-port junction circulators are joined together with the common junction matched to form a circulator with four accessible ports. Similarly, three, four, five, . . . , three-port circulators can be appropriately joined to form circulators with five, six, seven, . . . , accessible ports.

Let us now turn to some examples of specific applications of circulators. The nonswitchable circulators can be used to replace conventional TR (transmit-receive) tubes in microwave duplexing systems for simultaneous transmission and reception of energy. For a circulator with a $1 \rightarrow 2 \rightarrow 3 \rightarrow 1$ circulation pattern the transmitter is connected to port 1, the antenna to port 2, and the receiver to port 3. Thus, energy from the transmitter will go out to the antenna while a signal received by the antenna (simultaneously or otherwise) will be fed to the receiver. In this way there is little danger of the large transmitted power burning out the receiver detector in a rader set.

Circulators are also used to couple together circuit elements in two-terminal negative resistance, low-noise maser, parametric or tunnel diode amplifiers. For example, ports 1–3 in a $1 \rightarrow 2 \rightarrow 3 \rightarrow 1$ circulator are connected, respectively, to the pump, the cavity containing the maser material, and the load.

For high-power breakdown testing under short-circuit conditions, a circulator can be used to direct the reflected energy into a high-power termination.

A four-port switchable circulator can be used to switch a transmitter output between two antennas. In this case the transmitter is connected to port 1, the antennas to ports 2 and 4, and the receiver to port 3. A switchable circulator functioning as a SPDT (single-pole double-throw) switch can provide radar silence on a standby basis. Transmitter power is not turned off between transmission but is switched to a dummy load.

Pulse-latching digital circulators in either the differential phase shift or ferrite-loaded junction configuration can provide extremely fast switching speeds. In phased-array antennas switchable circulators can be used as digital phase shifters.

9.6.3 Application of Other Components

Variable attenuators and phase shifters can be constructed by using a centrally located axially oriented ferrite pencil in a rectangular waveguide with an external coil wound around it. In general, the attenuation and phase shift increase nonlinearly with coil current. Although they have many advantages over equivalent mechanical units, they are less accurate and more difficult to reset owing to the presence of ferrite hysteresis.

[35]For the reason behind this behavior, see the discussion in Sections 9.4 and 9.5.

A ferrite sample in a cavity can be used as a bandpass or band-elimination filter. The kind of filter action depends upon the geometrical configuration and the mode used. If the location of the passband is changeable by changing the field applied to the ferrite, the bandpass device can function as a scanning radar signal detector. In these cases single-crystal garnets are used because of their narrow resonance linewidth giving rise to the narrow passband. Small linewidth resulting in low spin wave excitation threshold also makes garnets useful in power limiters.

A YIG sphere is also used in a Gunn oscillator circuit for tuning purposes. The output frequency of the oscillator is then a function of the magnetic field applied to the garnet sample. Magnetostatic waves in YIG can be used for time delay. For microwave ICs, ferrite devices are fabricated in striplines, slot lines, and coplanar lines.

PROBLEMS

9.1 Show that the dc field orientations shown in Fig. 9.1 give low loss for energy propagated in the positive-y direction and high loss for energy propagated in the negative-y direction at ferromagnetic resonance.

9.2 Find and sketch the electric field E_z for the dielectric-loaded waveguide of Figs. 9.3 and 9.4.

9.3 Write down a set of phase shift equations for the four-port ring junction circulator of Fig. 9.12(b) similar to Eqs. (9.46) and (9.47) and show that it is not possible to obtain perfect circulation using this simple structure.

9.4 With reference to Fig. 9.14(a), it was stated that power entering terminal 1 will split between b and d and arrive at a and c in phase. Sum the phase shifts for the two paths and show that this is indeed the case.

9.5 (a) The scattering matrix of an isolator $[s]$ is given by

$$[s] = \begin{bmatrix} s_{11} & s_{12} \\ s_{21} & s_{22} \end{bmatrix}$$

write the expressions for b_1, b_2 in terms of a_1, a_2 and the components of $[s]$.

(b) Devise a set of experiments to measure s_{11}, s_{12}, s_{21}, and s_{22}. Would you expect $s_{11} = s_{22}$ and $s_{12} = s_{21}$ in general? Explain.

Superconducting Microwave Devices

The electrical resistivity of many metals and alloys drops abruptly to zero when they are cooled to a sufficiently low temperature T_c. Below this *critical temperature* the resistivity is identically zero[1] and the specimen is said to be in the superconducting state. Recently a number of microwave superconductor devices have appeared. These devices, as well as the related theories of superconductivity, will be discussed in this chapter.

10.1 SUPERCONDUCTIVITY

A superconductor is not just a perfect electrical conductor. It is also a perfect diamagnet at $T < T_c$. If it were just a perfect conductor, we would expect $\mathbf{E} = \mathbf{i}/\sigma$ to be zero for finite $\mathbf{i}$ when $\sigma = \infty$. It follows from Maxwell's equation that $d\mathbf{B}/dt = -\nabla \times \mathbf{E}$ would also be zero. Thus, this result predicts that the magnetic flux $\mathbf{B}$ through the metal could not be changed in cooling through the transition, contrary to the Meissner effect. It was observed by Meissner and Uchsenfeld[2] that when a specimen in a magnetic field is cooled through the transition temperature T_c, the magnetic flux originally present inside the specimen at $T > T_c$ is expelled from it when the temperature is lowered to a value less than T_c. In other words a superconductor at $T < T_c$ is a perfect conductor *and* a perfect diamagnet.

---

[1] J. File and R. G. Mills, *Phys. Rev. Lett.* **10**, 93 (1963).
[2] W. Meissner and R. Uchsenfeld, *Naturwiss.* **21**, 787 (1933).

Superconductivity occurs in many metallic elements, alloys, intermetallic compounds, and semiconductors.[3] Unfortunately their T_c extends only from about $10^{-2}°$K for some semiconductors to about 20°K for some compounds (e.g., $T_c \simeq 18°$K for Nb_3Sn, niobium tin). This means that a specimen must be immersed in liquid helium to take advantage of its superconducting properties, a substantial deterrent to widespread engineering applications.

10.1.1 Macroscopic Theory

Our understanding of the superconducting state is based on the laws governing thermodynamic transitions and the phenomenological equations of London and Landau-Ginsburg, as well as the quantum-mechanical theory of Bardeen, Cooper, and Schrieffer (BCS). A brief sketch of the BCS theory will be given in Section 10.1.2. First, however, let us discuss the macroscopic theories.

According to the Meissner effect, if any magnetic flux lines exist inside the sample at $T > T_c$, they will be expelled when the specimen is cooled through the transition temperature T_c. This effect is reversible in that the flux distribution inside the specimen will again be the same as before cooling through the transition if the temperature is raised once again above T_c. If a magnetic field H_a is applied parallel to the axis of a long specimen, the demagnetizing field inside the specimen will be negligible. In this case the Meissner effect can be conveniently expressed by the relation

$$B = \mu_0(H_a + M) = 0 \tag{10.1}$$

where B is the flux density inside the specimen at $T < T_c$. From this equation we find that the magnetization M due to the superconducting currents in the specimen is equal to $-H_a$. Actually, strictly speaking, Eq. (10.1) holds only when the Meissner effect is *complete*, i.e., when there is no flux penetration whatsoever into the specimen. However, according to experiments on thin superconducting films, there is *finite flux penetration* into the specimen. By modifying Ohm's law, London was able to develop an equation that satisfactorily accounts for finite flux penetration in superconductors.[4]

Instead of the usual Ohm's law, $\mathbf{i} = \sigma\mathbf{E}$ for ordinary conductors, London postulated that in a superconductor we have instead

$$\mathbf{i}_s = C\mathbf{A} \tag{10.2}$$

where C is a constant characteristic of a given material and $\mathbf{A}$ is the vector potential. Taking the curl of both sides of Eq. (10.2) and noting that $\mathbf{B} = \nabla \times \mathbf{A}$, we find

$$\nabla \times \mathbf{i}_s = C\mathbf{B} \tag{10.3}$$

---

[3]For tables of the properties of superconducting elements and compounds, see, e.g., C. Kittel, *Introduction to Solid State Physics*, Wiley, New York, 4th ed., 1971, p. 402.

[4]F. London and H. London, *Proc. R. Soc. London Ser. A* **149**, 72 (1935); *Physica* **2**, 341 (1935).

But one of Maxwell's equations states that for the static case,

$$\mathbf{V} \times \mathbf{B} = \mu_0 \mathbf{i}_s \qquad (10.4)$$

Taking the curl of both sides of Eq. (10.4) we have

$$-\nabla^2 \mathbf{B} = \mu_0 \mathbf{V} \times \mathbf{i}_s \qquad (10.5)$$

where we have noted that $\mathbf{V} \times \mathbf{V} \times \mathbf{B} = \mathbf{V}(\mathbf{V} \cdot \mathbf{B}) - \nabla^2 \mathbf{B}$ and $\mathbf{V} \cdot \mathbf{B} = \mathbf{0}$. Combining Eqs. (10.3) and (10.5) we finally obtain

$$\nabla^2 \mathbf{B} = -\mu_0 C \mathbf{B} \qquad (10.6)$$

The general solution of Eq. (10.6) for a semi-infinite superconductor occupying the space $x > 0$ is

$$B(x) = B(0) e^{-x/\lambda_L} \qquad (10.7)$$

where

$$\lambda_L^2 = -\frac{1}{\mu_0 C} \qquad (10.7a)$$

and $\mathbf{B}$ is assumed to lie in the y-z plane. Here λ_L is the length characterizing the rate of decay of the flux density from the value $B(0)$ at the surface to the value $B(x)$ at x. Accordingly, λ_L is known as the London penetration depth. Actually the penetration depth is not given by λ_L alone; the London equation (10.6) is somewhat simplified, as we shall discuss below.

It is interesting to note from Eq. (10.6) that $\mathbf{B}$ cannot be truly uniform inside a superconductor unless $\mathbf{B}$ itself is identically zero. This is because $\nabla^2 \mathbf{B} = 0$ if $\mathbf{B}$ is independent of the spatial coordinates whether $\mathbf{B}$ is zero or not.

If $\mathbf{B}$ is zero throughout the specimen but finite outside, then $\mathbf{B}$ must be a step function of x with the step located at $x = 0$. However, according to Eq. (2.58), the boundary condition on $\mathbf{B}$ at the 1 (air) to 2 (superconductor) interface is

$$\hat{\mathbf{x}} \times (\mathbf{B}_2 - \mathbf{B}_1) = \frac{\mathbf{K}}{\mu_0} \qquad (10.8)$$

where $\mathbf{K} = \mathbf{i} \, \Delta l$ is the surface linear current density (in amperes per meter) and i is the area current density (in amperes per squares meters). Δl then is the thickness of the layer through which the current is distributed. If $\mathbf{B}$ is a step function, $\mathbf{B}_2 - \mathbf{B}_1$ must be finite as $\Delta l \to 0$. Therefore $\mathbf{i} \to \infty$ for a finite $\mathbf{K}$. In other words the external magnetic flux density $\mathbf{B}_1$ is terminated at the superconductor surface by an infinitely thin current sheet of infinite area density. This situation is quite analogous to that at the surface of a waveguide wall of infinite conductivity except that the case in point is for a dc and not an r.f. magnetic field.

It remains unclear, however, as to whether λ_L can indeed attain zero value in a superconductor. An answer to this question can be obtained via the following simple observation. In general, the current density $\mathbf{i}$ in a conductor is given by

$$\mathbf{i}_s = n_s q \mathbf{v} \qquad (10.9)$$

where n_s is the density of carriers of charge q and $\mathbf{v}$ is their drift velocity. In the presence of a magnetic field $\mathbf{B} = \nabla \times \mathbf{A}$, $\mathbf{v}$ is related to the total momentum $\mathbf{p}$, mass m, and $\mathbf{A}$ by the relation[5]

$$\mathbf{p} = m\mathbf{v} + q\mathbf{A} \tag{10.10}$$

where the last term in the additional momentum due to the presence of the magnetic field. Solving for $\mathbf{v}$ and substituting the resulting expression into Eq. (10.9), we have

$$\mathbf{i} = \frac{n_s q}{m} \mathbf{p} - \frac{n_s q^2}{m} \mathbf{A} \tag{10.11}$$

It is interesting to note that Eq. (10.11) is consistent with the London equation (10.2) provided $\mathbf{p}$ is zero and

$$C = -\frac{n_s q^2}{m} \tag{10.12}$$

Combining Eqs. (10.7a) and (10.12), we finally obtain

$$\lambda_L^2 = \frac{m}{\mu_0 n_s q^2} \tag{10.13}$$

Since we do not expect m to be zero or $\mu_0 n_s q^2$ to approach infinity, λ_L for a superconductor should always be finite. Strictly speaking, then, the Meissner effect of flux exclusion is never truly complete in a superconductor. Just how nearly complete it is, of course, depends on the value of λ_L compared to the relevant dimensions of the sample.

Before we attempt a numerical estimate of λ_L, it is necessary to determine what values of m, n_s, and q to use. A clue to this comes from the requirement of zero total momentum p (due to drift velocity *and* magnetic field) mentioned in connection with Eq. (10.11) above. According to the quantum-mechanical theory of Bardeen, Cooper, and Schrieffer, loosely associated pairs of electrons are formed in the superconducting state.[6] The electrons in a given pair have momenta $\hbar\mathbf{k}$ and $-\hbar\mathbf{k}$ so that the net pair momentum is equal to zero. The formation energy for such pairs are provided by their interaction with the lattice, and the motion of superconducting electrons are correlated within a characteristic length known as the coherence length.[7] Without going into further details, it seems already evident that since the electron pairs behave more or less like a single particle, pair values should be used for m, n_s, and q in the calculation of λ_L given by Eq. (10.13). In other words $m = 2m_0$, $q = 2e$, and $n_s = n_0/2$ where

[5]For derivation of this equation, see, e.g., C. Kittel, *Introduction to Solid State Physics*, 4th ed., Addison-Wesley, Reading, MA, 1971, p. 727.

[6]J. Bardeen, L. N. Cooper, and J. R. Schrieffer, *Phys. Rev.* **106**, 162 (1957); **108**, 1175 (1957).

[7]Experimentally, it has been found that $M^\alpha T_c$ is constant, where M is the isotopic mass and $\alpha = \frac{1}{2}$ from the simple BCS theory neglecting Coulomb interaction between the electrons. Since T_c is dependent on M, we know that lattice vibrations, and therefore electron-lattice interactions must be responsible for superconductivity.

m_0 and e are the mass and charge of a single electron and n_0 is the free electron density of the sample.

For lead, a typical superconducting metal, $n_0 = 13.2 \times 10^{22}$ electrons/cm^3 and Eq. (10.13) gives a λ_L of 146 Å. Typical experimental penetration depths are in the order of hundreds of angstroms so that λ_L estimated from Eq. (10.13) gives the right order of magnitude. Since λ_L is typically small compared to the dimensions of a bulk superconductor, flux exclusion from the interior of the superconductor is nearly complete in that **B** is practically zero for all parts of the superconductor except for a region of a few hundred angstroms thick near the surface of the sample. However, for a thin film sample whose thickness is in the order of λ_L or smaller, the exclusion is far from complete and **B** may be quite nonuniform inside the sample.

Aside from the microwave applications to be discussed, it is obvious that a superconducting coil can be used to generate a large dc magnetic field by the application of a small voltage to its terminals, since the resistance of the superconductor is zero.[8] Such high magnetic fields can be used, for example, for plasma confinement in thermonuclear fusion research. However, it should be pointed out that the current density in the wire must not exceed a certain critical value lest the superconductivity be extinguished. This comes about because the current in the superconductor generates its own magnetic field, which in turn quenches the superconductivity. It is an experimental fact that a sufficiently strong magnetic field will destroy superconductivity. The threshold field H_c (whether applied or internally generated) for such destruction is a function of the temperature T. It is zero at $T = T_c$ and rises with decreasing temperature below T_c to a value of several hundred oersteds as $T \to 0$. H_c is always too low for a pure specimen of many materials for them to have any useful applications as superconducting coils. Such superconductors are known as type I superconductors.

Other superconductors, mostly alloys of transition metals with high normal state electrical resistivity, tend to have much higher H_c than those of type I (ten times as high, say). They are known as type II superconductors. These superconductors have superconducting properties up to an upper critical field H_{C2}. Between H_{C2} and the lower critical field H_{C1}, **B** inside the superconductor is not zero and the superconductor is threaded by flux lines and is said to be in the *vortex state*. In this state there is a large amount of magnetic hysteresis or flux pinning induced by mechanical treatment. Type II superconducting coils (of, say, Nb_3Sn) can supply dc fields of over 100 kOe.

10.1.2 Energy Gap and Electron Motion

To understand the microwave properties of a superconductor, it is first necessary to discuss the existence of an energy gap in a superconductor centered

---

[8]Current in the coil will not be infinite for a finite applied voltage, however, because of source and contact resistances.

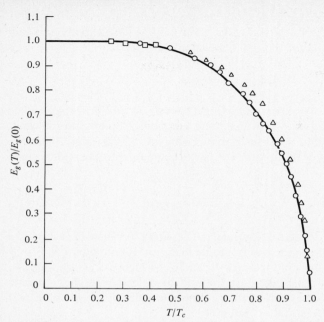

Figure 10.1 Normalized energy gap vs normalized temperature for superconductors. (Triangles: tin; circles: tantalum; squares: niobium; solid curve: BCS curve.)

about the Fermi level E_F of the normal state. This energy gap is of an entirely different nature than those in semiconductors or insulators and is typically very small. E_g at $T = 0°K$ ranges from about 1.5 to 30×10^{-4} eV depending on the material; for Pb, e.g., $T_c = 7.2°K$ and $E_g(0) \simeq 6 \times 10^{-4}$ eV. This energy gap comes about because the superconducting state, consisting of loosely coupled electron pairs, is more ordered than the normal state. Since a finite energy is needed to break up a pair of electrons in the superconducting state, we would expect a superconducting electron to gain an energy of the order of the net pair coupling energy below T_c to become truly free and normal again. At $T = T_c$ the energy gap is zero as expected, since at the transition temperature the coupling energy is nullified by the thermal agitation energy kT_c giving rise to zero net coupling energy. As T decreases, thermal agitation also decreases, and the net coupling energy and energy gap should therefore increase with decreasing T. This supposition is consistent with the results of BCS theory and with experiment, as depicted in Fig. 10.1.[9] At $T = 0$ there are no electrons above the energy gap and all electrons are of the superconducting type. As T increases from 0, electron pairs are broken up and excited across the energy gap by thermal agitation. At $T = T_c$ all electrons become normal, and for $0 < T < T_c$ an admixture of normal and superconducting electrons coexists in the sample.

[9]P. Townsend and J. Sutton, *Phys. Rev.* **128**, 591 (1962).

Although electrons are lossless in the superconducting state, they have finite mass and inertia. Thus, the equation of motion for the n_s superconducting electrons per unit volume for the one dimensional case is

$$n_s m_s \frac{d^2 x_s}{dt^2} = n_s f(t) \tag{10.14}$$

where x_s is the position of a superconducting electron pair and m_s its mass while $f(t)$ is the force acting on m_s, which may be a function of time t. On the other hand the motion of normal electrons is damped by lattice scattering. Thus, the corresponding equation of motion for the n_n normal electrons per unit volume is

$$n_n m_n \frac{d^2 x_n}{dt^2} + n_n \alpha_n \frac{dx_n}{dt} = n_n f(t) \tag{10.15}$$

where α is a damping constant. Equations (10.14) and (10.15) are based on the supposition that the superconductor is infinite in extent. If boundary conditions are imposed, the equivalent inductances L_s and L_n, which will be obtained in Eqs. (10.16) and (10.17), must take into account the difference in characteristic decay distance for **B**, λ_s (the London decay length) for the superconductor, and δ (the skin depth) for the normal conductor.

To better understand the behavior of superconductors at microwave frequencies, let us find the electric equivalent circuit that obeys differential equations of the same form as Eqs. (10.14) and (10.15). If we think of m (inertia) as corresponding to L, α (damping) to resistance, x to i, and $f(t)$ to $v'(t)$, these equations can be immediately transformed [with $m_s \rightarrow L_s$, $x_s \rightarrow i_s$, $f(t) \rightarrow v'(t)$] to

$$L_s \frac{d^2 i_s}{dt^2} = v'(t) \tag{10.16}$$

and [with $m_n \rightarrow L_n$, $\alpha_n \rightarrow R_n$, $x_n \rightarrow i_n$, $f(t) \rightarrow v'(t)$] to

$$L_n \frac{d^2 i_n}{dt^2} + R_n \frac{di_n}{dt} = v'(t) \tag{10.17}$$

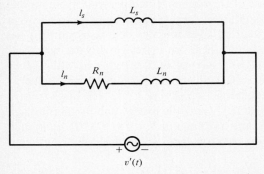

Figure 10.2   Equivalent circuit of a superconductor.

The corresponding circuit, as shown in Fig. 10.2, is a parallel combination of L_s, and L_n and R_n in series, across a source $v'(t)$.

Referring to Fig. 10.2, we see that for direct current ωL_s and ωL_n go to zero and R_n is shorted out by ωL_s. Thus, for $T < T_c$, although there are both superconducting and normal electrons present, the resistivity (α impedance seen by v') is identically zero. If ω is finite, however, R_n will no longer be shorted and the resistivity of the sample will be finite even for $T < T_c$. With this observation, we are close to an understanding of the behavior of superconductors at microwave frequencies (Section 10.1.4).

10.1.3 Surface Impedance

To derive an expression for the surface impedance, it is first necessary to derive the wave equation based on the two-fluid model, i.e., a model based on the simultaneous existence of superconducting and normal electrons. This can be accomplished by combining Eq. (10.2) with Maxwell's equations as follows. Taking the curl of both sides of

$$\mathbf{V} \times \mathbf{H} = \mathbf{i} + \frac{\partial \mathbf{D}}{\partial t} \tag{2.2}$$

we have

$$\nabla^2 \mathbf{H} = -\mathbf{V} \times \mathbf{i} - \epsilon_0 \frac{\partial \mathbf{E}}{\partial t} \tag{10.18}$$

where we have used Eqs. (2.4) and (2.9) and the vector identity $\mathbf{V} \times \mathbf{V} \times \mathbf{H} = \mathbf{V}(\mathbf{V}\mathbf{H}) - \nabla^2\mathbf{H}$. Now combining Eq. (10.18) with Eq. (2.1), we have

$$\nabla^2 \mathbf{H} = -\mathbf{V} \times \mathbf{i} + \mu_0\epsilon_0 \frac{\partial^2 \mathbf{H}}{\partial t^2} \tag{10.19}$$

where we have also used Eq. (2.8).

Based on the two-fluid model, we can write

$$\mathbf{i} = \mathbf{i}_s + \mathbf{i}_n \tag{10.20}$$

where $\mathbf{i}_s$ and $\mathbf{i}_n$ are, respectively, superconducting and normal current densities. Taking the curl of both sides of Eq. (10.20) and using Eqs. (10.3) and (10.7a) and Ohm's law, $\mathbf{i}_n = \sigma\mathbf{E}$, we have

$$\mathbf{V} \times \mathbf{i} = -\frac{1}{\lambda_L^2} \mathbf{H} - \mu_0\sigma \frac{\partial \mathbf{H}}{\partial t} \tag{10.21}$$

where we have used Eqs. (2.1) and (2.8) once again. Finally, combining Eqs. (10.19) and (10.21) we obtain

$$\nabla^2 \mathbf{H} = \frac{1}{\lambda_L^2} \mathbf{H} + \mu_0\sigma \frac{\partial \mathbf{H}}{\partial t} + \mu_0\epsilon_0 \frac{\partial^2 \mathbf{H}}{\partial t^2} \tag{10.22}$$

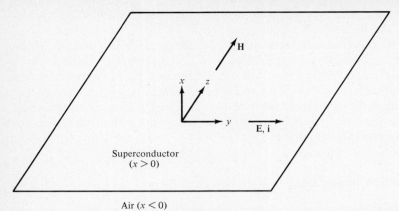

Figure 10.3   Coordinate system for superconductor surface impedance calculation.

Let us apply Eq. (10.22) to the case of a semi-infinite superconductor occupying all space with $x > 0$, as shown in Fig. 10.3. If $\mathbf{E}$ and $\mathbf{i}$ are in the y direction, $\mathbf{H}$ is along the z direction. With only an x dependence and an $e^{j\omega t}$ time dependence, Eq. (10.22) becomes

$$\frac{\partial^2 H_z}{\partial x^2} = \frac{1}{\lambda_L^2} H_z + \frac{j}{\delta^2} H_z - \omega^2 \mu_0 \epsilon_0 H_z \qquad (10.23)$$

where $\delta = 1/\sqrt{\pi f \mu_0 \sigma}$ is the skin depth for normal conductor. At microwave frequencies, the last term on the right-hand side of Eq. (10.23), whose origin can be traced back to the displacement current in Maxwell's equations, can be easily neglected. Substituting $H_z = H_0 e^{-kx}$ into Eq. (10.23), we find

$$k = \frac{1}{\lambda_L}\left[\left(\frac{\alpha+1}{2}\right)^{1/2} + j\left(\frac{\alpha-1}{2}\right)^{1/2}\right] \qquad (10.24)$$

where
$$\alpha = \left[1 + 4\left(\frac{\lambda}{\delta}\right)^4\right]^{1/2} \qquad (10.25)$$

As $\omega \to 0$, $\delta \to \infty$. It follows from Eq. (10.25) that $\alpha \to 1$ and from (10.24) that $k \to 1/\lambda_L$, as expected.

It is convenient at this point to define a measurable quantity known as the surface impedance Z:

$$Z = E_0 \bigg/ \int_0^\infty i_y \, dx \qquad (10.26)$$

where E_0 is the magnitude of the y-directed electric field at the surface of the superconductor and $\int_0^\infty i_y \, dx$ is the total current carried by the superconductor. The integral $\int_0^\infty i_y \, dx$ can be evaluated with the aid of the Maxwell equation (2.2), which in one dimension becomes

$$-\frac{\partial H_z}{\partial x} = i_y \qquad (10.27)$$

Thus,

$$\int_0^\infty i_y \, dx = \int_0^\infty k H_0 e^{-kx} \, dx = H_0 \tag{10.28}$$

Therefore, the expression for Z simply becomes

$$Z = \frac{E_0}{H_0} \tag{10.29}$$

To complete the calculation, it is necessary to express E_0 in terms of H_0 using the other Maxwell's equation [Eq. (2.1)]:

$$\frac{\partial E_y}{\partial x} = -j\omega\mu_0 H_z \tag{10.30}$$

Noting that $E_y = E_0 e^{-kx}$ and $H_z = H_0 e^{-kx}$ in Eq. (10.30), we find:

$$E_0 = \frac{j\omega\mu_0 H_0}{k} \tag{10.31}$$

Substituting Eq. (10.31) into Eq. (10.29) and using Eq. (10.24), we finally obtain an expression for the surface impedance Z:

$$Z = R + jX = \frac{j\omega\mu_0}{k} = \omega\mu_0\lambda_L\left(\frac{\alpha - 1}{2\alpha^2}\right)^{1/2} + j\omega\mu_0\lambda_L\left(\frac{\alpha + 1}{2\alpha^2}\right)^{1/2} \tag{10.32}$$

If a superconducting resonant cavity is used, R can be measured precisely by measurement of Q for the cavity while X can be determined by noting the resonance frequency shift from the expected value based on cavity geometry.

10.1.4 Microwave Properties

The existence of an energy gap has a profound effect on the surface resistance of superconductors at microwave frequencies. For photon energy less than the energy gap, the surface resistance of a superconductor vanishes at $T = 0$ even at finite frequencies, as all electrons are then at the superconducting state. For photon energies higher than the energy gap, transitions to the normal state occur, giving rise to a finite surface resistance. This behavior is shown in Fig. 10.4. Again, in the case of lead, the critical frequency $f_c = E_g(0)/h = 17.5$ GHz, which is in the microwave band. As T is increased above absolute zero, the decrease in gap energy (see Fig. 10.1) is reflected in the downward shift of the knee points in the surface resistance vs frequency curve of Fig. 10.4.[10] In addition, it is seen that the resistivity below the knee no longer vanishes for $\omega > 0$. For $\omega > 0$, ωL_s of Fig. 10.2 is no longer zero and R_n is no longer shorted as for direct current. In other words, the inertia of the superconducting electrons prevents them from completely screening the electric field, and so thermally excited normal electrons can absorb energy.

[10]M. A. Biondi and M. P. Garfunkel, *Phys. Rev. Lett.* **2**, 143 (1959).

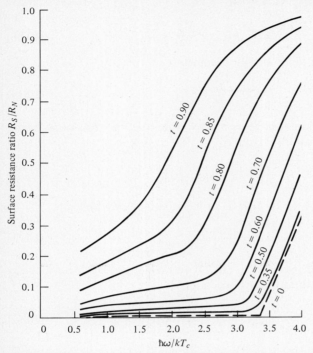

Figure 10.4 Normalized surface resistance of aluminum vs (normalized) frequency in the microwave band. $t = T/T_c$.

10.2 QUANTUM-MECHANICAL TUNNELING

Perhaps the most useful superconducting phenomenon in microwave applications is Josephson tunneling. For example, a Josephson junction can be used as the active, nonlinear element in a parametric amplifier. Before we discuss the Josephson tunneling phenomenon, it is advisable for us first to review the case of normal quantum-mechanical tunneling

10.2.1 Normal Tunneling

Figure 10.5 shows the potential energy as a function of position for a one-dimensional barrier of finite width d and height U_0. The Schrödinger equations for the three regions of Fig. 10.5 are

$$\frac{d^2\psi_1}{dx^2} + \frac{2m}{\hbar^2} E\psi_1 = 0 \tag{10.33}$$

$$\frac{d^2\psi_2}{dx^2} + \frac{2m}{\hbar^2} (E - U_0)\psi_2 = 0 \tag{10.34}$$

$$\frac{d^2\psi_3}{dx^2} + \frac{2m}{\hbar^2} E\psi_3 = 0 \tag{10.35}$$

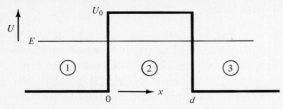

Figure 10.5   Normal tunneling through a potential barrier.

where the ψs are the wave functions and $E\ (< U_0)$ is the energy of the electron penetrating the barrier. Note that, in contrast to the classical case, there is a finite probability for the electron to go from region 1 to region 3 even though its energy is less than the barrier height.

Solutions of Eqs. (10.33)–(10.35) are of the form

$$\psi_1 = Ae^{i\alpha x} + A'e^{-i\alpha x} \tag{10.36}$$

$$\psi_2 = Be^{\beta x} + B'e^{-\beta x} \tag{10.37}$$

$$\psi_3 = Ce^{i\alpha x} \tag{10.38}$$

where

$$\alpha^2 = \frac{2mE}{\hbar^2} \tag{10.39}$$

$$\beta^2 = \frac{2m(U_0 - E)}{\hbar^2} \tag{10.40}$$

The constants A, A', B, B', and C can be evaluated by matching ψ and $d\psi/dx$ at $x = 0, d$. After these constants have been determined and appropriate substitutions are made in Eqs. (10.36)–(10.38), the transmission coefficient can be determined:

$$\frac{C^2}{A^2} = \frac{4(U_0/E - 1)}{(U_0/E - 2)^2 \sinh^2(d/2d_0) + 4(U_0/E - 1)\cosh^2(d/2d_0)} \tag{10.41}$$

where

$$d_0 = \frac{1}{2\beta} = \frac{\hbar}{2}\sqrt{\frac{1}{2m(U_0 - E)}} \tag{10.42}$$

If $d \gg 2d_0$, $\sinh d/2d_0 \simeq \cosh d/2d_0 \simeq \frac{1}{2}e^{d/2d_0}$ and Eq. (10.41) becomes

$$\frac{C^2}{A^2} = 16\left(\frac{E}{U_0}\right)\left(1 - \frac{E}{U_0}\right)e^{-d/d_0} \tag{10.43}$$

Clearly, C^2/A^2 decreases rapidly with increasing d/d_0. Typically, $U_0 - E$ is on the order of 1 eV. If the free electron mass is used, Eq. (10.42) gives $d_0 \simeq 1$ Å. Thus, d must be very small for appreciable tunneling to occur. Fortunately, on the other hand, C^2/A^2 decreases relatively slowly with decreasing E/U_0 according to Eq. (10.43), indicating that particles at all energy levels $E\ (< U_0)$ can participate in the tunneling process. It follows that, despite the e^{-d/d_0} factor, the tunneling current can be quite respectable. This clearly has important practical implications in microwave devices employing the tunneling phenomenon.

10.2.2 Metal-Insulator-Metal Junctions

We now apply the concept of tunneling to a metal-insulator-metal (MIM) sandwich. Depending on whether the metals are in the normal or superconducting states, the behavior of these junctions can be classified into three types: (1) normal conductor–insulator–normal conductor (NIN), (2) normal conductor–insulator–superconductor (NIS), and (3) superconductor-insulator-superconductor (SIS). In any case the metals on either side of the insulator need not be the same.

Let us first consider the case of tunneling through an MIM sandwich when both metals are still in their normal states. In the interest of generality, let the two metals be dissimilar. In thermal equilibrium the Fermi levels of the two metals are at the same level, as depicted in the energy E vs density-of-states N_1, N_2 diagram of Fig. 10.6(a). If a voltage V of polarity shown in Fig. 10.6(b) is applied to the sandwich, all the electrons in the left metal will have an energy eV electron volts higher than those in the right metal. This leads to a relative shift in the Fermi levels, as shown in Fig. 10.6(b), and a consequent net tunneling current across the insulator.

Let us now calculate the tunneling current as a function of applied voltage V. The number of electrons within an energy interval dE that will move from left to right should be proportional to the number of occupied states on the left, that is, to

$$N_1(E - eV)P_{\mathrm{FD}}(E - eV)\,dE$$

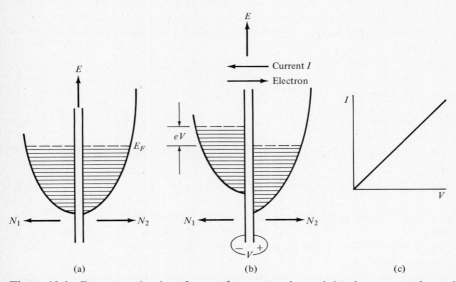

Figure 10.6 Energy vs density of states for a normal metal–insulator–normal metal junction at $T = 0$ **(a)** at thermal equilibrium and **(b)** with an applied voltage V. **(c)** Junction I-V characteristics.

and to the number of unoccupied states on the right, or to

$$N_2(E)[1 - P_{FD}(E)]$$

where N_1 and N_2 are densities of states and P_{FD} is the Fermi-Dirac probability of occupation of these states. It follows that the current flow from left to right is given by

$$I_{l \to r} \sim P_{12}(E)N_1(E - eV)P_{FD}(E - eV)N_2(E)[1 - P_{FD}(E)] \, dE \quad (10.44)$$

where $P_{12}(E)$ is the probability of transition across the barrier at an energy E. Similarly, the current flow from right to left is given by

$$I_{r \to l} \sim P_{21}(E)N_1(E - eV)[1 - P_{FD}(E - eV)]N_2(E)P_{FD}(E) \, dE \quad (10.45)$$

Making the reasonable assumption that an electron has just as much chance to tunnel from left to right as to tunnel in the opposite direction, i.e., $P_{21} = P_{12}$, we obtain the net current by combining Eqs. (10.44) and (10.45) and integrating over all energies:

$$I \sim \int P_{12}(E)N_1(E - eV)N_2(E)[P_{FD}(E - eV) - P_{FD}(E)] \, dE \quad (10.46)$$

For small applied voltages it seems reasonable to assume that $P_{12}(E)$ is nearly independent of energy, and if the densities of states are slowly varying functions of energy near the Fermi level, we can use their values at the Fermi level. Thus,

$$N_1(E - eV) \simeq N_1(E) \simeq N_1(0)$$

and

$$N_2(E) \simeq N_2(0)$$

Accordingly, Eq. (10.46) simplifies to

$$I = -eAP_{12}N_1(0)N_2(0) \int [P_{FD}(E - eV) - P_{FD}(E)] \, dE \quad (10.47)$$

where e is the electron charge and A is the cross-sectional area of the sandwich. The minus sign has been inserted to indicate that the net electron current flows from right to left.

The quantity $P_{FD}(E - eV) - P_{FD}(E)$ can be expanded by means of the Taylor series to yield

$$[P_{FD}(E - eV) - P_{FD}(E)] = P_{FD}(E) - eV \frac{\partial P_{FD}(E)}{\partial E} - P_{FD}(E) - \cdots$$

For small V this quantity can thus be approximated by $-eV \, \partial P_{FD}(E)/\delta E$. If the temperature is not high, $-eV \, \partial P_{FD}(E)/\partial E$ can in turn be approximated by a δ function, resulting in

$$I = -[eAP_{12}N_1(0)N_2(0)]V \quad (10.48)$$

Clearly, an NIN junction is ohmic, i.e., $I \sim V$, as shown in Fig. 10.6(c).

Next, let us consider the NIS case in which one of the metals is superconducting. It turns out that its I-V characteristic, unlike that of the NIN sandwich, is not linear for $|V| > 0$. Indeed, the current remains quite small until

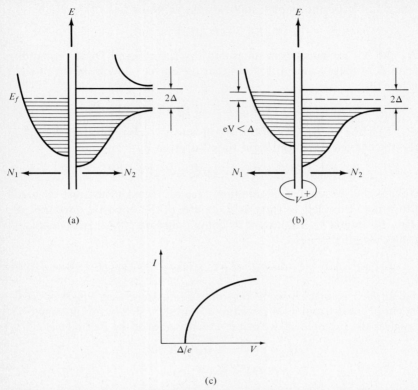

(a)   (b)

(c)

Figure 10.7   Energy vs density of states for a normal metal insulator. E_F denotes the Fermi level. Superconductor junction at $T = 0$ **(a)** at thermal equilibrium and **(b)** with $V < \Delta/e$. **(c)** Junction I-V characteristics.

V reaches some critical value V_c, at which point I increases rapidly with V. The explanation for this behavior lies in the existence of an energy gap 2Δ and singularities in the density of states curve at the boundaries of the gap, as shown in Fig. 10.7(a).

The sketch of the density of states curve for the superconductor in Fig. 10.7(a) is in accord with the BCS theory of superconductivity, which gives[11]

$$N_S(E) = N_N(E) \frac{E}{\sqrt{E^2 - \Delta^2}} \tag{10.49}$$

where N_S and N_N are the densities of states for a given metal in the superconducting and normal states, respectively, while E is measured from the Fermi level located at the center of the gap. Clearly, a singularity occurs for N_S at the band edges, i.e., at $E = \pm\Delta$.

[11]J. Bardeen, L. N. Cooper, and J. R. Schrieffer, *Phys. Rev.* **108**, 1175 (1957).

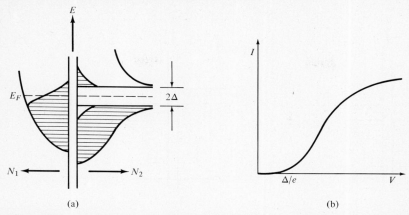

Figure 10.8 **(a)** Energy vs density of states for a normal metal–insulator–superconductor junction at $T = 0$. **(b)** Junction I-V characteristics.

At $T=0$ all states in the normal metal (metal 1) are filled up to E_F, whereas in the superconductor (metal 2) all states below the gap are filled and all states above the gap are empty. Of course, the Fermi levels of metals 1 and 2 coincide in thermal equilibrium. When a voltage $V < \Delta/e$ is applied [see Fig. 10.7(b)], electrons in metal 1 have no access to empty states in metal 2. Therefore, no current flows for $V < \Delta/e$ [see Fig. 10.7(c)]. At $V = \Delta/e$ the current rises rapidly because of the accessibility of empty states to electrons of metal 1 and the large density of states at the upper edge of the band gap. For $V > \Delta/e$ the I-V curve bends toward the V axis, as shown in Fig. 10.7(c), because of the decrease in the density of states as V departs somewhat from Δ/e. I never drops to zero, however, because $2\Delta \ll E_F$; the scale used in Fig. 10.7(a) and 10.7(b) is highly distorted.

For $T > 0$ some electrons in metal 1 have energies greater than $E_F + \Delta$ even in thermal equilibrium and some normal electrons exist above the gap in metal 2, as shown in Fig. 10.8(a). Now a very small voltage is sufficient for the commencement of current flow, but any appreciable increase in I must still occur around $V = \Delta/e$, as shown in Fig. 10.8(b).

Consider next the SIS sandwich. For the case where the two metals are the same, the energy diagram in thermal equilibrium is as shown in Fig. 10.9(a). At $T = 0$ all energy levels are filled up to $E_F - \Delta$ as shown. If $V < 2\Delta/e$, $I = 0$ as the electrons below the band gap in superconductor 1 have no access to empty states in metal 2. At $V = 2\Delta/e$, however, a sudden rise in current occurs as shown in Fig. 10.9(c) since electrons in superconductor 1 just below the band gap suddenly gain access to the enormous number of states just above the band gap in superconductor 2 [see Fig. 10.9(b)]. For $T > 0$ some states above the gap will be filled even in thermal equilibrium. As a consequence, the I-V curve will be somewhat rounded off.

If the superconductors in an SIS sandwich are not of the same type, then the energy diagram at $T > 0$ is as shown in Fig. 10.10(a). It is left as an exercise

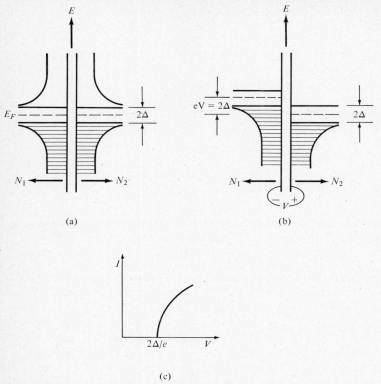

(a)   (b)

(c)

Figure 10.9   Energy vs density of states for a superconductor-insulator-superconductor junction made of identical superconductors at $T = 0$ **(a)** at thermal equilibrium and **(b)** with $V = 2\Delta/e$. **(c)** Junction I-V characteristics.

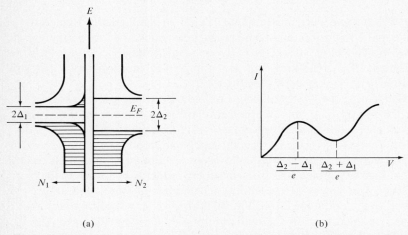

(a)   (b)

Figure 10.10   **(a)** Energy vs density of states for a superconductor-insulator-superconductor junction made of dissimilar superconductors at $T > 0$. **(b)** Junction I-V characteristics.

for the reader (using the procedure outlined above) to show that the I-V curve at $T > 0$ is as shown in Fig. 10.10(b). Note, in particular, from Fig. 10.10(b) that between $V = (\Delta_2 - \Delta_1)/e$ and $V = (\Delta_2 + \Delta_1)/e$ the differential resistance is negative, indicating the possibility of using this type of SIS sandwich as an energy source.

10.3 JOSEPHSON EFFECT

In Section 10.2 single-electron tunneling in NIN, NIS, and SIS sandwiches were discussed. Most of the microwave superconductor devices to date, however, utilize the Josephson effect associated with the tunneling of superconducting electron pairs in an SIS sandwich. The effects of pair tunneling are quite different from those of single-particle tunneling. Indeed, they include these puzzling phenomena:

1. In the *dc Josephson effect*, a dc current flows across the junction in the absence of any electric or magnetic field.
2. In the *ac Josephson effect*, a dc voltage applied across the junction causes r.f. current oscillations at frequency f across the junction. Conversely, in the presence of microwave radiation at frequency f, dc supercurrents (i.e., the transfer of Cooper pairs across the barrier) can occur provided the potential difference is such that energy can be conserved by absorption or stimulated emission of one or more photons. For a particular value of microwave power, a specimen would spontaneously jump onto a constant voltage region in the I-V characteristic. Thus, it behaves like an ideal zero-impedance voltage source powered by microwave energy.
3. If a small r.f. field is applied to an unbiased Josephson junction, oscillation in phase occurs. In this case the barrier behaves with respect to supercurrents exactly like a variable inductor which can serve as the coupling element in microwave parametric amplifiers.
4. A dc magnetic field applied through a two-junction superconducting circuit causes the maximum supercurrent to show interference effects as a function of magnetic field. This effect can be utilized to build sensitive magnetometers.

In the discussion to follow, we shall first give a simplified quantum-mechanical treatment of phenomena 1 and 2. Then we proceed to a more extensive discussion of phenomenon 3 with illustrative applications to microwave parametric amplifier circuits. Item 4 will not be discussed, for it is not within the scope of microwave magnetics.

10.3.1 DC Josephson Effect

Consider an unbiased SIS sandwich. Let ψ_1 be the probability amplitude of electron pairs on one side of the junction, and let ψ_2 be that on the other side. The time-dependent Schrödinger equation $H\psi = i\hbar\,\partial\psi/\partial t$ applied to the two regions becomes

$$\hbar T\psi_2 = i\hbar\frac{\partial\psi_1}{\partial t} \tag{10.50}$$

and

$$\hbar T\psi_1 = i\hbar\frac{\partial\psi_2}{\partial t} \tag{10.51}$$

Here $\hbar T$ represents the "transfer interaction Hamiltonian" across the insulator, a measure of the leakage of ψ_1 into region 2 and of ψ_2 into region 1. Thus, $|\psi_1|^2$ represents the density of superconducting electron pairs that managed to cross the barrier from region 1 to region 2, and $|\psi_2|^2$ is that from region 2 to region 1. Clearly, the value of T is dependent on the thickness of the insulator; as barrier thickness approaches infinity, $T \to 0$ and pair tunneling ceases.

To solve Eqs. (10.50) and (10.51), let

$$\psi_1 = n_1^{1/2}e^{i\theta_1} \tag{10.52}$$

and

$$\psi_2 = n_2^{1/2}e^{i\theta_2} \tag{10.53}$$

Substituting Eqs. (10.52) and (10.53) for ψ_1 and ψ_2 into Eqs. (10.50) and (10.51), we have

$$\tfrac{1}{2}n_1^{-1/2}e^{i\theta_1}\frac{\partial n_1}{\partial t} + in_1^{1/2}e^{i\theta_1}\frac{\partial\theta_1}{\partial t} = -iTn_2^{1/2}e^{i\theta_2} \tag{10.54}$$

$$\tfrac{1}{2}n_2^{-1/2}e^{i\theta_2}\frac{\partial n_2}{\partial t} + in_2^{1/2}e^{i\theta_2}\frac{\partial\theta_2}{\partial t} = -iTn_1^{1/2}e^{i\theta_1} \tag{10.55}$$

Multiplying Eq. (10.54) by $n_1^{1/2}e^{-i\theta_1}$ and Eq. (10.55) by $n_2^{1/2}e^{-i\theta_2}$, we obtain for $\delta = \theta_2 - \theta_1$

$$\frac{1}{2}\frac{\partial n_1}{\partial t} + in_1\frac{\partial\theta_1}{\partial t} = -iT(n_1 n_2)^{1/2}e^{i\delta} \tag{10.56}$$

and

$$\frac{1}{2}\frac{\partial n_2}{\partial t} + in_2\frac{\partial\theta_2}{\partial t} = -iT(n_1 n_2)^{1/2}e^{-i\delta} \tag{10.57}$$

Equating real and imaginary parts of both sides of Eq. (10.56) and proceeding similarly for Eq. (10.57), we find

$$\frac{\partial n_1}{\partial t} = 2T(n_1 n_2)^{1/2}\sin\delta \tag{10.58}$$

$$\frac{\partial n_2}{\partial t} = -2T(n_1 n_2)^{1/2}\sin\delta \tag{10.59}$$

and

$$\frac{\partial\theta_1}{\partial t} = -T\left(\frac{n_2}{n_1}\right)^{1/2}\cos\delta \tag{10.60}$$

$$\frac{\partial\theta_2}{\partial t} = -T\left(\frac{n_1}{n_2}\right)^{1/2}\cos\delta \tag{10.61}$$

If the superconductors on both sides are of the same type, $n_1 = n_2$. Equations (10.58) and (10.59) show that

$$\frac{\partial n_2}{\partial t} = -\frac{\partial n_1}{\partial t} \tag{10.62}$$

Similarly, letting $n_1 = n_2$ in Eqs. (10.60) and (10.61) shows that

$$\frac{\partial(\theta_2 - \theta_1)_0}{\partial t} = \frac{\partial \delta_0}{\partial t} = 0 \tag{10.63}$$

The conclusions expressed by Eqs. (10.62) and (10.63) are most interesting. First, from Eq. (10.63), the phase difference between ψ_1 and ψ_2 (the so-called order parameters) is independent of time. Second, the current flow from region 1 to region 2 is proportional to $\partial n_2/\partial t$ or, equivalently, to $-\partial n_1/\partial t$. Note in particular that, since $\partial n_2/\partial t = -\partial n_1/\partial t$, pairs are crossing in one direction only; i.e., the pair gain per unit time in region 2 is exactly equal to the pair loss per unit time in region 1. We can therefore expect a finite current flow at zero applied voltage. Indeed, the magnitude J_0 of the current flow density, according to Eqs. (10.58) and (10.59), should be given by

$$j_0 = J_0 \sin \delta_0 = J_0 \sin(\theta_2 - \theta_1)_0 \tag{10.64}$$

where $$J_0 = e^* 2T(n_1 n_2)^{1/2} \tag{10.65}$$

where $e^* = 2e$ is the charge of the electron pair. The supercurrent density i_0 can thus take on various values between $-J_0$ and $+J_0$ depending on the phase difference $\theta_2 - \theta_1$ of the order parameters.

10.3.2 AC Josephson Effect

If a voltage, which may be a function of time, is applied across the junction, an electron pair will experience a potential energy difference of 2 eV across the barrier. Equivalently, we can say that an electron pair is at a potential $-eV$ on one side and $+eV$ on the other side of the junction. Accordingly, the left-hand side of Eqs. (10.50) and (10.51) must be modified by the addition of $-eV\psi_1$ and $+eV\psi_2$ terms, respectively, to yield

$$\hbar T\psi_2 - eV\psi_1 = i\hbar \frac{\partial \psi_1}{\partial t} \tag{10.66}$$

$$\hbar T\psi_1 + eV\psi_2 = i\hbar \frac{\partial \psi_2}{\partial t} \tag{10.67}$$

We now proceed to solve Eqs. (10.66) and (10.67) in exactly the same way as we solved Eqs. (10.50) and (10.51). The result is

$$\frac{\partial n_2}{\partial t} = -\frac{\partial n_1}{\partial t} \tag{10.68}$$

and $$\frac{\partial(\theta_2 - \theta_1)_1}{\partial t} = \frac{\partial \delta_1}{\partial t} = -\frac{2\,eV}{\hbar} \tag{10.69}$$

which of course reduces to Eq. (10.63) when $V \to 0$. The corresponding super-current density is given by

$$j_1 = J_0 \sin \delta_1 \tag{10.70}$$

Although this expression for j_1 is of the same form as that for j_0 for the $V = 0$ case, given by Eq. (10.64), δ_1 here is determined by Eq. (10.69) rather than by Eq. (10.63) for δ_0. If $V = V_0$ is independent of t, then integration of Eq. (10.69) yields

$$\int_{\delta_1(0)}^{\delta_1} d\delta_1 = \frac{-2\,eV_0}{h} \int_0^t dt \quad \text{or} \quad \delta_1 = \delta_1(0) - \frac{2\,eV_0}{h} t \tag{10.71}$$

where $\delta_1(0) = \delta_1$ at $t = 0$.

Substituting Eq. (10.71) for δ_1 into Eq. (10.70), we obtain

$$
\begin{aligned}
j_1 &= J_0 \sin\left[\delta_1(0) - \frac{2\,eV_0}{h} t \right] \\
&= -J_0 \sin[\omega_1 t - \delta_1(0)]
\end{aligned}
\tag{10.72}
$$

Thus, we observe the remarkable effect of a dc voltage V_0 applied across the junction, giving rise to an oscillating current of frequency $f_1 = 2\,eV_0/h = 4.8 \times 10^{14} V_0$. This is known as the ac Josephson effect. Note that a voltage V_0 of the order of $10\ \mu V$ can generate r.f. oscillations in the microwave range.

10.3.3 Josephson Junction as Variable Inductance

From Eq. (10.70) we obtain, by differentiation with respect to t,

$$\frac{dj_1}{dt} = J_0 \cos \delta_1 \frac{d\delta_1}{dt} \tag{10.73}$$

Combining this equation with Eq. (10.69), we have

$$\frac{dj_1}{dt} = -\left(\frac{2eJ_0}{h} \cos \delta_1 \right) V \tag{10.74}$$

Rearranging we have

$$V = -\left(\frac{h \sec \delta_1}{2eJ_0} \right) \frac{dj_1}{dt} = -l \frac{dj_1}{dt} \tag{10.75}$$

Where l clearly has the dimension of an inductance (per unit cross-sectional area).

If the junction is unbiased (with respect to direct current) and $V = V_m \cos \omega t$, integration of Eq. (10.69) gives

$$\delta_1 = \delta_1(0) - \frac{2\,eV_m}{h\omega} \sin \omega t \tag{10.76}$$

It then follows from Eq. (10.75) that

$$l = \frac{\hbar}{2eJ_0} \sec\left[\delta_1(0) - \frac{2\,eV_m}{\hbar\omega} \sin \omega t \right] \tag{10.77}$$

For very small V_m, Eq. (10.77) simplifies to

$$l \simeq l_0 - l_1 \sin \omega t \tag{10.78}$$

where
$$l_0 = \sec \delta_1(0)$$

$$l_1 = \frac{2\,eV_m}{\hbar\omega} \sec \delta_1(0) \tan \delta_1(0) \tag{10.79}$$

provided $\delta_1(0) \neq n\pi \neq (2n - 1)\pi/2$. Thus, we see that an unbiased Josephson junction will act as a series combination of a constant and a time-varying inductance. This variable inductance can be employed as the coupling element in a microwave parametric amplifier (see Section 10.4).

10.4 JOSEPHSON JUNCTION PARAMETRIC AMPLIFIERS

Equation (10.78) shows that an unbiased Josephson junction can function as a time-varying inductance to be used, say, as the coupling element in a parametric amplifier. Because of the extremely high Q of such an inductor, the noise figure of a superamp (superconducting parametric amplifier) should yield an extraordinary low noise figure. In this section we shall discuss the operation of superamps, devices that have appeared in the last few years.

Figure 10.11 shows four of a series of 80 unbiased junctions incorporated into the upper conductor of a broadband 50-Ω microstrip transmission line. They were fabricated by an intersection-scratch technique.[12] Using a fine diamond tool, a fine groove was scribed along the axis of a highly polished fused quartz substrate. A 1000-Å tin film was then evaporated onto the grooved surface of this substrate. Next the substrate was scribed with 80 lighter grooves perpendicular to the axial tin-covered groove. These lighter grooves removed the tin everywhere along their paths except at the intersection with the axial groove. As seen from Fig. 10.11, this process produces a series of microbridges (~ 1500 Å long $\times$ 2500 Å wide). Spacing between adjacent microbridges or junctions was chosen to be 4 μm so that the total length of the 80 series-connected junctions is about 320 μm. This represents only $\frac{1}{45}$ of a wavelength of the strip line [highly polished fused quartz ($\epsilon_r \simeq 4$) between tin and lead conductors] at 10 GHz, ensuring that microwave currents flow through all junctions essentially in phase. Note that the evaporated lead-ground plane served as both an electric and a magnetic shield. Termination is accomplished by means of an open circuit located at $\lambda/4$ after the junction. This is equivalent to placing the microbridge assembly across the transmission line at

[12]R. Y. Chiao, M. J. Feldman, H. Ohta, and P. T. Parrish, *J. Phys. Appl. Paris* **9**, 183 (1974).

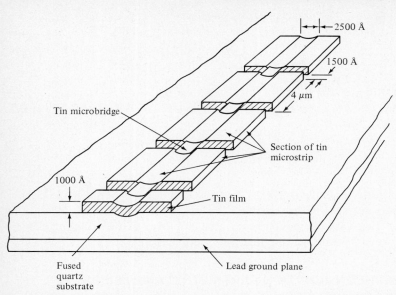

Figure 10.11 Diagram (not drawn to scale) showing sections of a microstrip connected by microbridges across which Josephson junction tunneling takes place.

10 GHz and at its odd harmonics.[13] To obtain maximum power transfer between the strip line and the microbridge assembly, the number of micro-bridges is adjusted so that the series impedance of the assembly is equal to the 50-Ω, or characteristic, impedance of the line.

A gain of 12 dB at 10 GHz was obtained when the assembly was used as the time-varying element in a parametric amplifier with a bandwidth of 1 GHz. As expected, the noise temperature ($<20°$K) was extremely low.[14] The supa-ramp was operated in the "doubly degenerate" mode with pump, signal, and idle frequencies closely and equally spaced, according to the relation[15,16]

$$2\omega_p = \omega_s + \omega_i \qquad (10.80)$$

A 9-mm (33-GHz) suparamp using the microbridge assembly described for the 10-GHz suparamp has also been constructed.[17] A bandwidth of 34 GHz and a suparamp contribution of only $20° \pm 10°$K to the noise temperature was obtained.

---

[13]Actually, a strip line without a termination is not truly open circuited as it is embedded in free space with an impedance of 377 Ω. If the line is shorted at $\lambda/2$ after the microbridges, they will be connected in shunt across the strip line at 10 GHz and all its harmonics.

[14]P. T. Parrish and R. Y. Chiao, *Appl. Phys. Lett.* **25**, 627 (1974).

[15]A. S. Clorfein, *Appl. Phys. Lett* **4**, 131 (1964); *Proc. IEEE* **52**, 844 (1964); *Proc. IEEE* **53**, 388 (1965).

[16]H. Zimmer, *Appl. Phys. Lett.* **10**, 193 (1967).

[17]R. Y. Chiao and P. T. Parrish, "Operation of the Suparamp at 33 GHz" (unpublished).

To understand how a microbridge can serve as a Josephson junction, let us return to an inspection of Fig. 10.11. According to this figure, the upper conductor is divided into a number of complete sections connected by "bridges" between them. Since the width of the "bridges" (~ 2500 Å) is rather narrow compared to the width of a complete section of the upper conductor, we can imagine the microwave current lines to be squeezed as it passes through the bridges. Furthermore, since the length of a bridge (~ 1500 Å) is also very small, there can nevertheless be a reasonable probability that the current (or pairs) will cross this narrow bridge. Thus, we can expect the occurrence of the Josephson effect even in the absence of any apparent junction.

A detailed theory of the suparamp can be found in the literature.[18]

PROBLEMS

10.1 Since superconducting electrons are expected to collide with the lattice, with normal electrons, and with each other, why is their motion, unlike that of normal electrons, undamped [see Eqs. (10.14) and (10.15)]?

10.2 Calculate and sketch the ratio R/R_n from Fig. 10.2. Try to reconcile your results with that given in Eq. (10.32).

10.3 Derive Eq. (10.41).

10.4 Justify the shape of the I-V characteristic of Fig. 10.10(b).

[18]M. J. Feldman, P. T. Parrish, and R. Y. Chiao, *J. Appl. Phys.* **46**, 4031 (1975).

Appendix: Gaussian and MKS Units

A.1 GAUSSIAN UNITS

In the Gaussian system of units Maxwell's equations are

$$\mathbf{\nabla} \times \mathbf{E} = -\frac{1}{c}\frac{\partial \mathbf{B}}{\partial t} \tag{A.1}$$

$$\mathbf{\nabla} \times \mathbf{H} = \frac{1}{c}\left(4\pi i + \frac{\partial \mathbf{D}}{\partial t}\right) \tag{A.2}$$

$$\mathbf{\nabla} \cdot \mathbf{D} = 4\pi\rho \tag{A.3}$$

$$\mathbf{\nabla} \cdot \mathbf{B} = 0 \tag{A.4}$$

Here all electric quantities, i.e., $\mathbf{D}$, $\mathbf{E}$, and ρ, are in electrostatic units, while all magnetic quantities, i.e., $\mathbf{B}$, $\mathbf{H}$, and i, are in electromagnetic units; $c = 3 \times 10^{10}$ cm/s is the velocity of light in free space. In addition to Eqs. (A.1)–(A.4), there are two constitutive equations:

$$\mathbf{B} = \mathbf{H} + 4\pi\mathbf{M} \tag{A.5}$$

$$\mathbf{D} = \mathbf{E} + 4\pi\mathbf{P} \tag{A.6}$$

Equations (A.5) and (A.6) can be transformed to

$$\mathbf{B} = (1 + 4\pi\|\mathbf{\chi}\|)\mathbf{H} \tag{A.7}$$

$$\mathbf{D} = (1 + 4\pi\|\mathbf{\chi}_e\|)\mathbf{E} \tag{A.8}$$

where $\|\mathbf{\chi}\|\mathbf{H} = \mathbf{M}$ and $\|\mathbf{\chi}_e\|\mathbf{E} = \mathbf{P}$.

The magnetic and electric susceptibilities $\|\mathbf{\chi}\|$ and $\|\mathbf{\chi}_e\|$ are in general tensors. However, for ferromagnetics, $\|\mathbf{\chi}\|$ is a tensor but χ_e is a scalar. Thus, $\|\mathbf{\mu}\| = 1 + 4\pi\|\mathbf{\chi}\|$

is a permeability tensor while $\epsilon = 1 + 4\pi\chi_e$ is the dielectric constant of a ferromagnet. Note that in Eq. (A.5) B and $4\pi M$ are in gauss while H is in oersteds.

A.2 MKS UNITS

In MKS, or practical, units, Maxwell's equations are

$$\mathbf{V} \times \mathbf{E} = \frac{\partial \mathbf{B}}{\partial t} \tag{A.9}$$

$$\mathbf{V} \times \mathbf{H} = i + \frac{\partial \mathbf{D}}{\partial t} \tag{A.10}$$

$$\mathbf{V} \cdot \mathbf{D} = \rho \tag{A.11}$$

$$\mathbf{V} \cdot \mathbf{B} = 0 \tag{A.12}$$

with the constitutive equations

$$\mathbf{B} = \mu_0(\mathbf{H} + \mathbf{M}) \tag{A.13}$$

$$\mathbf{D} = \epsilon_0 \mathbf{E} + \mathbf{P} \tag{A.14}$$

where　$\mathbf{E}$ = electric intensity, V/m
　　　　$\mathbf{D}$ = electric displacement, C/m^2
　　　　ρ = charge density, C/m^3
　　　　i = current density, A/m^2
　　　　$\mathbf{H}$ = magnetic intensity, A/m
　　　　$\mathbf{M}$ = magnetization, A/m
　　　　$\mathbf{B}$ = magnetic flux density, Wb/m^2

and all lengths are in meters; time is in seconds.

The permeability for free space μ_0 is equal to $4\pi \times 10^{-7}$ F/m and the dielectric constant for free space $\epsilon_0 = (1/36\pi) \times 10^{-9}$ H/m. It follows that $1/\sqrt{\mu_0\epsilon_0}$ is equal to 3×10^8 m/s, the velocity of light in free space, and $\sqrt{\mu_0/\epsilon_0} = 377 \, \Omega$ is the impedance of free space.

In the field of magnetism $4\pi M$ and H are usually given in gauss and oersteds, respectively—i.e., in electromagnetic units. However, since MKS units are used throughout this book, it is appropriate to cite the pertinent conversion factors here:

$$B \text{ (G)} \times 10^{-4} = B \text{ (Wb/m}^2\text{)}$$

$$4\pi M \text{ (G)} \times 79.5 = M \text{ (A/m)}$$

$$H \text{ (Oe)} \times 79.5 = H \text{ (A/m)}$$

Analogous to Eqs. (A.5) and (A.6), Eqs. (A.13) and (A.14) can be changed to

$$\mathbf{B} = \mu_0(1 + \|\chi\|)\mathbf{H} \tag{A.15}$$

$$\mathbf{D} = \epsilon_0(1 + \|\chi_e\|)\mathbf{E} \tag{A.16}$$

where $\|\chi\|$ and $\|\chi_e\|$, the magnetic and electric susceptibilities, are generally tensors, although in a ferromagnet χ_e is a scalar. $\|\mu_r\| = 1 + \|\chi\|$ is defined as the permeability tensor relative to free space, while $\epsilon_r = 1 + \chi_e$ is the dielectric constant of a ferromagnet relative to free space.

Since the ratio M/H in both Eqs. (A.7) and (A.15) are dimensionless quantities, we see that

$$4\pi\|\chi\| \text{ (Gaussian)} = \|\chi\| \text{ (MKS)} \tag{A.17}$$

Frequently in the literature on ferrites, $4\pi\|\chi\|$ (Gaussian) is calculated from H and $4\pi M$ given in oersteds and gauss, respectively. From relation (A.17), we observe that the calculated value of $4\pi\|\chi\|$ (Gaussian) can be used for $\|\chi\|$ in Maxwell's equations expressed in MKS units.

Additional References

For those who wish to pursue certain topics covered in this book, consult the following appropriate level of readable sources:

Microwaves, Waveguides, and Cavities

S. Ramo, J. R. Whinnery, and T. Van Duzer, *Fields and Waves in Communication and Electronics*, Wiley, New York, 1965.

IEEE. Trans. Microwave Theory Tech.

Magnetism and Magnetic Materials

B. D. Cullity, *Introduction to Magnetic Materials*, Addison-Wesley, Reading, MA, 1972.

A. H. Morrish, *The Physical Principles of Magnetism*, Wiley, New York, 1965.

S. Chikaazumi, *Physics of Magnetism*, Wiley, New York, 1959.

IEEE. Trans. Magnetics, especially yearly Intermag Conf. Proceedings.

J. Appl. Phys., especially yearly Conference on Magnetism Magnetic Materials Proceedings.

Resonance

C. P. Slichter, *Principles of Magnetic Resonance*, Harper & Row, New York, 1963.

R. F. Soohoo, *Magnetic Thin Films*, Harper & Row, New York, 1965.

Microwave Ferrite Devices

R. F. Soohoo, *Theory and Application of Ferrites*, Prentice-Hall, Englewood Cliffs, NJ, 1960.

IEEE. Trans. Microwave Theory Tech.

IEEE. Trans. Magnetics.

Superconducting Devices

L. Solymar, *Superconductive Tunnelling and Applications*, Wiley-Interscience, New York, 1972.

Author Index

Subject Index

Absorption curve, 146
Alloys, 85
Angular momentum, 64
  orbital, 69, 72, 73, 84, 85
  spin, 69, 72
  total, 69, 72
Anisotropic media, 182–188
  Faraday rotation in, 185–186
  propagation in, 182–185, 187–188
Anisotropic ordering, 86
Anisotropy
  antiferromagnetic, 179
  due to magnetostriction, 86
  easy direction of, 84
  induced, 85–86
  surface, 85–86
  unidirectional, 86, 92
Anisotropy constants, 83–84
Anisotropy energy, 83–86, 92, 104, 170
  definition of, 83
  dependence on strain of, 92
  within domain wall, 94
Anisotropy field, 170–172, 205
Antiferromagnetic coupling, 98
Antiferromagnetic layer, 86
Antiferromagnetic materials, 97, 194
Antiferromagnetic resonance, 179–180
  equation of motion of, 180
  frequency of, 86, 180
Antiferromagnetism, 81, 95–97

Antiferromagnets, resistivities of, 97
Atomic clock, 148, 161
Atomic number, 63
Attenuation constant, 6, 12

Backward wave oscillator, 127
Bessel functions, 21–22
Bethe-Schwinger perturbation relation, 144
Bloch equations, 119, 122–123, 130–131
Bloch wall, 94–95
Bohr magneton, 71–72, 114, 116
Boltzmann constant, 68
Boltzmann distribution, 71, 134, 137, 159
Brillouin function, 72

Classical mechanics, 60, 64
Coaxial lines, 32–35
Convection current, 9
Coplanar lines, 225
Coulomb energy, 79, 80–81
Crystal field, 73, 90–91
Crystal periodicity, 83
Crystal stress, 92
Crystal strain, 92
Cubic close-packed fcc lattice, 98
Curie constant, 72, 82
Curie law, 72, 78, 104
Curie temperature, 74, 75, 79, 82
  and exchange integral, 82